# THE ROTHSCHILDS AND THE GOLD RUSH

## Benjamin Davidson and Heinrich Schliemann in California, 1851–52

# THE ROTHSCHILDS AND THE GOLD RUSH

## Benjamin Davidson and Heinrich Schliemann in California, 1851–52

### Giles Constable

American Philosophical Society Press
Philadelphia •'2015

Transactions of the
American Philosophical Society
Held at Philadelphia
For Promoting Useful Knowledge
Volume 105, Part 4

ISBN: 978-1-60618-054-9

U.S. ISSN: 0065-9746

**Library of Congress Cataloging-in-Publication Data**

Constable, Giles.
  The Rothschilds and the Gold Rush : Benjamin Davidson and Heinrich Schliemann in California 1851-52 / Giles Constable.
    pages cm. — (Transactions of the American Philosophical Society held at Philadelphia for promoting useful knowledge, ISSN 0065-9746 ; volume 105, part 4)
  Includes bibliographical references and index.
  ISBN 978-1-60618-054-9 (alkaline paper)
  1. California—Gold discoveries. 2. Gold mines and mining—California—History—19th century. 3. Business enterprises—California—San Francisco—History—19th century. 4. Davidson, Benjamin, 1823-1878. 5. Schliemann, Heinrich, 1822-1890. 6. Rothschild family. 7. Bankers—England—London—History—19th century. 8. Banks and banking—England—London—History—19th century. I. Title.
  F865.C75 2015
  979.4'04—dc23
                          2015019816

# Contents

# Introduction

In the middle of the nineteenth century, in response to the discovery of gold, tens of thousands prospective miners came to California from all over the world. There, they hoped to change the trajectory of their lives and the lives of their families within a few months or years. These discoveries and the subsequent migrations influenced the lives of citizens of many nations, the history of the American nation, and its newly acquired continental empire in the most direct and dramatic ways.

Over the last century and a half, the same iconic images have come to characterize the California Gold Rush. Rough men with beards bent over pans of gravel in rushing streams. In the background loomed the snow-covered peaks of the sierra. Sometimes the images included tents and a burro. The emphasis was on the individual (or a small group of men) caught in a cycle of endless physical labor in a harsh unforgiving landscape. The many accounts of the miners that survive support these images, at least in the first years of the gold rush.

The story of mining gold in California turned out to be long—California was a major gold producer fifty years later—and complicated. Mining evolved in its techniques and demands. Larger units of labor and capital succeeded individuals and small groups, as mining followed a natural path to economies of scale and new technologies. The pick, pan, and shovel were replaced by machines, investment capital, large labor forces, quartz mining deep underground, and stock available to the public. In this way, absentee investors had the opportunity to invest in mining enterprises far distant from their homes and businesses.

The transition to these more complex dimensions of the gold rush coincided with the rise of San Francisco as a center of economic enterprise. A village of about 1,000 at the time of the gold discoveries, San Francisco grew at an astonishing rate, commensurate with its location as the new destination of hundreds and soon tens of thousands. In addition to the gateway to the gold fields, it quickly became the center of supplies for the distant mining camps. As trade in commodities surged, San Francisco made the transition from a town of many commercial transactions to a city of professionals. Indeed, by the spring of 1851, arrivals in San Francisco could describe the place as a town growing into a city, characterized by problems and opportunities associated with urban growth everywhere. Fueled by the flow of gold from the mining camps, San Francisco telescoped a century of growth into the single year of 1849. By 1850, its population had risen to 20,000; by 1852, 36,000.

With the backing of major European financial houses, the purchase and remittance of California gold became a major business. Rothschild provided its agent, Benjamin Davidson, with a credit of £40,000, which was renewed monthly and could be drawn on houses in London, Paris, Frankfurt, Naples, Vienna, or New York. This array of cities identifies the range of the house of Rothschild. For his part, Davidson made monthly shipments of $150,000 to $300,000 (often nearer $100,000). In order to increase his harvest of gold, in September 1851, Davidson appointed Schliemann his agent in Sacramento. The two remained in partnership in the gold dust business for eight months.

Professor Constable's essay analyzes the stresses and strains in these partnerships that reflected the intensely competitive nature of the business. From the beginning, Davidson and Schliemann had frequent disagreements. They bickered, among other issues, over commissions, shipping arrangements, and assaying charges. Although neither fully trusted the other, they were joined in the gold dust business on a large scale.

The profitability of the gold dust business—as evidenced by the large European houses engaged in it—meant a high degree of competition. This competition was reflected in the rise of banks in San Francisco.

Davidson's bank, supported by Rothschild funds, was one of the first and certainly the strongest. As early as 1849, it was described as the only bank in San Francisco with a capital of $100,000. By 1853, San Francisco was home to nineteen banks, with capital many times as large. The competition between and among banks for the gold dust business was intense.

As gold made its way from California to Europe, so pressure for performance flowed in the opposite direction. As the competition intensified, so did the demands on the agents. By comparison with London or Paris (or even New York), the economy of San Francisco was chaotic. There were endless ways to make money in the city, each accompanied by parallel risks. Professor Constable observes that the Rothschilds were instinctively entrepreneurial and could not resist San Francisco's many opportunities for profit. At the same time as the Rothschild offices in London wanted Davidson to maximize his profits, they also demanded he minimize his risks. The correspondence between the agents and the offices in London show the pressures from both sides. That a letter from San Francisco took two months to reach London and a reply the same, mixed with the shifting market to create endless opportunities for misunderstandings, especially when letters crossed or conditions changed. And, in the background lay the gold market in England and Europe, the final arbiter in these complicated transactions.

The Davidson/Schliemann partnership ended in April 1852, when Schliemann hurriedly departed Sacramento. He left behind several questions, and Davidson stepped forward to tie up the loose ends of the partnership. When this business collaboration ended, a fascinating window on the early entrepreneurial history of California also closed.

Malcolm J. Rohrbough
Professor Emeritus
Department of History
University of Iowa

# Acknowledgments

Thanks for help in preparation of this manuscript are owed to Melanie Aspey and Victor Gray (London), Stefanie A. H. Kennell (Vancouver), Mary Morganti (San Francisco), John Mayo (University of the Western Indies), Peter Blodgett (Los Angeles), Kevin Morse (Sacramento), Malcolm J. Rohrbough (Scituate), Pat Johnson (Sacramento), and others who are named in the notes. The author gratefully acknowledges financial support from the Mellon Foundation, which enabled him to travel to Athens, London, and California to consult relevant sources.

# Illustrations

Cover photo: *San Francisco*, by S. F. Marryat. London: M. N. Hanhart Chromo Lith. Impt. [ca. 1850?]. Reproduced from Popular Graphic Arts Collection, Prints and Photographs Division, Library of Congress.

Figure 1. Benjamin Davidson in San Francisco. From an undated photograph in the Rothschild Archives.    7

Figure 2. Heinrich Schliemann in 1861. From a photograph in the Gennadius Library, Athens.    8

Figure 3. Sacramento City (showing the Tehama Block on the northeast corner of J and Front Streets). Reproduced from William Redmond Ryan, *Personal Adventures in Upper and Lower California in 1848–9*, 2 vols. (London, 1850), I, 163.    10

Figure 4. The Tehama Block Building was built in 1850 and torn down in 1851, when it was replaced with a brick building. Courtesy of the Sacramento County Historical Society.    11

Figure 5. The Tehama Block Building in a photograph of 1851. Schliemann's bank may have been located in the building on the extreme left. Courtesy of the Sacramento County Historical Society.    12

Figure 6. The Tehama Block in 1852 showing the premises of Page, Bacon & Co, Grim & Rumler, G. Frank Smith, E. and R. K. Swift, Adams & Co, Read & Co, and Wells Fargo & Co. From the *Sacramento Pictorial Union* (January 1853). Courtesy Center for Sacramento History.    13

# 1

# Arriving in California

In the minds of most people the Gold Rush summons up a picture of the thousands of prospectors and miners who flocked to California after the discovery of gold there in 1848, especially those who came in 1849, the so-called "Forty-niners." It has been estimated that the total number of migrants coming to California in 1849 exceeded 20,000 and that this figure doubled the following year and rose to a total of 250,000 by 1853. Not all of them went to the gold fields, however. They included many merchants, shopkeepers, bankers, doctors, clergy, craftsmen of various sorts, keepers of hotels and boarding houses, both men and women, not to mention less desirable types who lived off the prospectors. Many of these settled in the towns, especially San Francisco and Sacramento, which grew rapidly in this period. Though it is impossible to give precise figures, it is probable that a greater proportion of these service-providers, as they may be called, prospered more than the prospectors, many of whom failed in their endeavors and went home, or to Australia and Canada after the discovery of gold there.

Among these were two young men who laid the bases of their fortunes in California: the British banker Benjamin Davidson and the German merchant (and later archaeologist) Heinrich Schliemann, who were associated for a few months in 1851–52. Davidson was the agent in San Francisco for N. M. Rothschild and Sons, and Schliemann was a banker in Sacramento. So far as is known they had not met before, though both of them were in St. Petersburg in 1847–48, and they probably never saw each other again after 1852. During the last seven months of Schliemann's time in California, however, he had an active business relationship with Davidson, and their letters throw light not only on their own activities but also on the early history of California, particularly the economics of the gold rush.

The principal sources on Davidson are his letters in the Rothschild Archive in London.[1] There is also some important material in the archives of the California Historical Society in San Francisco and among the family papers preserved by his descendants, which include a record book of his elder brother David and the diploma appointing Davidson consul of Sardinia (later Italy) in San Francisco.[2] For Schliemann's career at this time the main sources are the letter-book of copies of his letters to Davidson, which is preserved among his papers in the Gennadius Library in Athens,[3] and material, much of it in contemporary newspapers, in the Sacramento City Archives and in the California State Library in Sacramento. The extensive secondary literature on Schliemann, aside from a few articles, is primarily concerned with his later life and archaeological work, as are his own autobiographical writings, which are notably unreliable. His so-called journal recounting his first visit to America is inaccurate, if not downright misleading, on many points. It is uncertain whether he was purposely mendacious or a master of self-deception.[4]

Davidson's letters are preserved both in the originals, written by himself or one of his clerks, and in copies marked "Duplicate," some of which are in German, or in French if they were sent to the Paris branch of Rothschild, and differ considerably from the originals (see Appendix). To carry a letter between San Francisco and London took approximately two months, which was a long time in a rapidly shifting market and afforded many openings for misunderstandings when letters crossed or conditions changed. Most of the letters went via Panama, being carried overland across the isthmus, but some went via New York and others round Cape Horn. Letters written at different times were often sent in batches by the same courier, as the opportunity offered, and were filed together in London, though some letters, of a more private nature, were filed separately and others, still unfound, may survive in different files in the Rothschild Archive. Most of the letters dealt with the technical details of shipments of gold and the sale of drafts transferring money from one place to another, but they also covered a wide variety of topics of general interest.[5]

Schliemann's letters are known from his letter-book, which is in itself an object of interest. It is a quarto volume, measuring twenty-two by twenty-six centimeters and consisting of 670 (plus one unnumbered) pages of special absorbent paper, bound in leather-backed marbled boards.[6] After a letter was written it was placed under one of the pages, insulated with blotting-paper, and pressed together, resulting in a copy of the letter, positive on the recto and negative on the verso.[7] In places the ink has faded and the paper rotted (especially around Schliemann's signature, which is written in almost solid ink), but the result is equivalent to reading the original letters. The register contains a total of 329 letters, of which 262 were addressed to Davidson—sometimes two or three a day—between 19 October 1851 and 6 April 1852. They were written either by Schliemann himself or by one of his two clerks, A. K. Grim, a native of Cleveland, and Miguel de Satrustegui, who came from San Sebastian, Spain. The majority of the letters to Davidson are in English, but a few are in German, Schliemann's native language, and other letters in the book are in Russian, Spanish, and Dutch. Even Schliemann's detractors admit that he was a remarkable linguist.[8] The letters to Davidson and the accompanying shipments of gold were carried from Sacramento to San Francisco down the Sacramento River by the overnight boats, which went every day except Sunday.

Of the two men, Davidson is the less well known and therefore needs a fuller introduction here. He was born on 12 January 1823, the fourth child of eight—seven sons and one daughter—of Meyer (Mayer) Davidson and his wife, Jesse. Meyer was born in Berlin in about 1785, came to England in 1803, and from about 1810 was associated in business with Nathan Mayer Rothschild, with whom he played an important part in financing the British troops in Europe in 1814–16.[9] In 1816 he married Jesse Cohen, the daughter of Levy Barent Cohen (1747–1808), a general merchant and dealer in diamonds.[10] Her sisters married Nathan Mayer Rothschild, Samuel Moses Samuel, and Moses Montefiore.[11] Meyer Davidson was therefore the brother-in-law of N. M. Rothschild, and his son Benjamin was first cousin to Rothschild's sons Lionel, Anthony, and Mayer, who jointly ran the London house after their father's death in 1836, and of Nathaniel, who worked in Paris with his uncle Baron James de Rothschild, Nathan's brother.[12] Through his mother Benjamin was related to some of the most prominent Jewish families in England.

Nothing is known about Davidson's early life aside from a reference in a letter dated 30 May 1851 to his having worked for the gas company (presumably as a young man in England) for an annual salary of £75. From 1847, and probably earlier, when he was in Paris, he worked for the Rothschilds, who liked to keep their business within the family.[13] In 1847, when he was twenty-five years old, he was sent to St. Petersburg, probably to establish a bank there, because he took with him £250,000 in gold, which was almost lost when one of his carriages fell into the Rhine at Cologne.[14] He stayed in St. Petersburg from at least 13 March until 9 November 1847 and made many important contacts there, but he failed to establish a bank, perhaps (the Rothschilds believed) because he was a Jew. He was back in England before the end of the year, and on 17 December he left for Valparaiso in Chile, where he stayed from at least 28 February 1848 until 28 June 1849 and was occupied with the Rothschild interests in quicksilver.[15]

An account of Davidson's departure from Valparaiso for California is given in a letter written to the Rothschilds on 12 November 1849 by his elder brother Lionel, who was the Rothschild agent in Mexico.[16]

> What I have written to my brother is this:—that as *you, he* and *I* seem all agreed on one point, viz. the inutility of his remaining in Valparaiso, doing nothing, my advice to him is to pack his traps as soon as the state of his pending business enables him to do so and to start—for California!—You will probably have already received various vague reports of the wonderful discoveries lately made in that district of "placenes de oro"—which are not exactly gold mines, but large districts in which the soil is impregnated with gold dust: and from which the precious metal is obtained with little trouble and at no expense beyond that of the labour of washing. . . . This has produced quite a social revolution in these parts.

Lionel advised his brother to go there and promised to make himself responsible to the Rothschilds that "they will not object to the expenses of his journey."[17] This important letter shows that Davidson was not sent to California by Rothschild, as has been said, but went there on the advice of his brother and at his own initiative.[18] This may account in part for the Rothschilds' concern about Davidson's activities in San Francisco, which will be discussed later. In any case, the gold rush was on, and Davidson was part of it. He was in fact a forty-niner, since he arrived in San Francisco on 18 August 1849 (Figure 1).

Schliemann also came to California of his own initiative. He was born in 1822, a year before Davidson, and was the fifth child of Adolph Schliemann, a Protestant minister. At the age of thirteen, according to his own account (which cannot be fully trusted) he was apprenticed to a grocer. He then worked for B. H. Schroeder and Co. in Amsterdam and was sent by them in 1846 or 1847 to St. Petersburg, where he spent five years (Figure 2). He at first dealt in indigo and later established, as he put it, "a mercantile house on my own account," dealing in various commodities, while still representing Schroeder.[19] His younger brother Ludwig (Louis) went to the United States in 1849 and settled in Sacramento. He wrote at least two letters from there to Schliemann, one dated

**Figure 1.** Benjamin Davidson in San Francisco.

From an undated photograph in the Rothschild Archives.

**Figure 2.** Heinrich Schliemann in 1861.
From a photograph in the Gennadius Library, Athens.

25 September 1849 and another dated 27 March 1850 (with a postscript, in German, dated 29 March), in which he described the banking business and the search for gold:

> The great rush is now for the Yuba, hundreds are daily starting, men averaging there now from 1½ ounces to 4 ounces pr. day, and those who possess a quicksilver machine can realize fully the double amount.

He went on to describe the booming real estate business in Sacramento, his own investments in land and commerce, and (in the postscript) how three "fellows" who lodged with him had broken into his strongbox and stolen $120.[20] Ludwig died of typhus on 21 May 1850, leaving an estate worth 7,000 thaler,[21] and Schliemann decided to go to California to arrange his brother's burial and claim his estate and also, probably, encouraged by Ludwig's glowing description of Sacramento, to try his own luck, because he clearly brought some capital with him. He arrived in San Francisco on the steamer *Oregon* on 3 April 1851.[22] After going to Sacramento to settle his brother's affairs, he returned briefly to San Francisco and on 9 April went back to Sacramento, where he established a banking house on the northeast corner of J and Front Streets, a short distance from where the boats to and from San Francisco docked (Figure 3).[23]

The only source for this stage of Schliemann's career is his journal, in which the first entry from Sacramento is dated 26 April and gives a description of the town and its commercial activities. Far from being a wealthy city, as he had expected, Schliemann wrote, money was tight, interest rates high, and speculation, especially in real estate, rife: "In no country of the world have I found so much selfishness and such immense love of money as in this Eldorado." From 14–19 May he went by steamer to Maryville, where he visited the Yuba diggings, and then to Parkis Bar, Nevada City, and the Gold Run and Grass Valley, where he saw several quartz mills. On 26 May he was in San Francisco and visited the Sonoma and Napa Valleys before returning to Sacramento and settling down to work (Figures 4–6).

**Figure 3.** Sacramento City (showing the Tehama Block on the northeast corner of J and Front Streets).
Reproduced from William Redmond Ryan, *Personal Adventures in Upper and Lower California in 1848–9*, 2 vols. (London, 1850), I, 163.

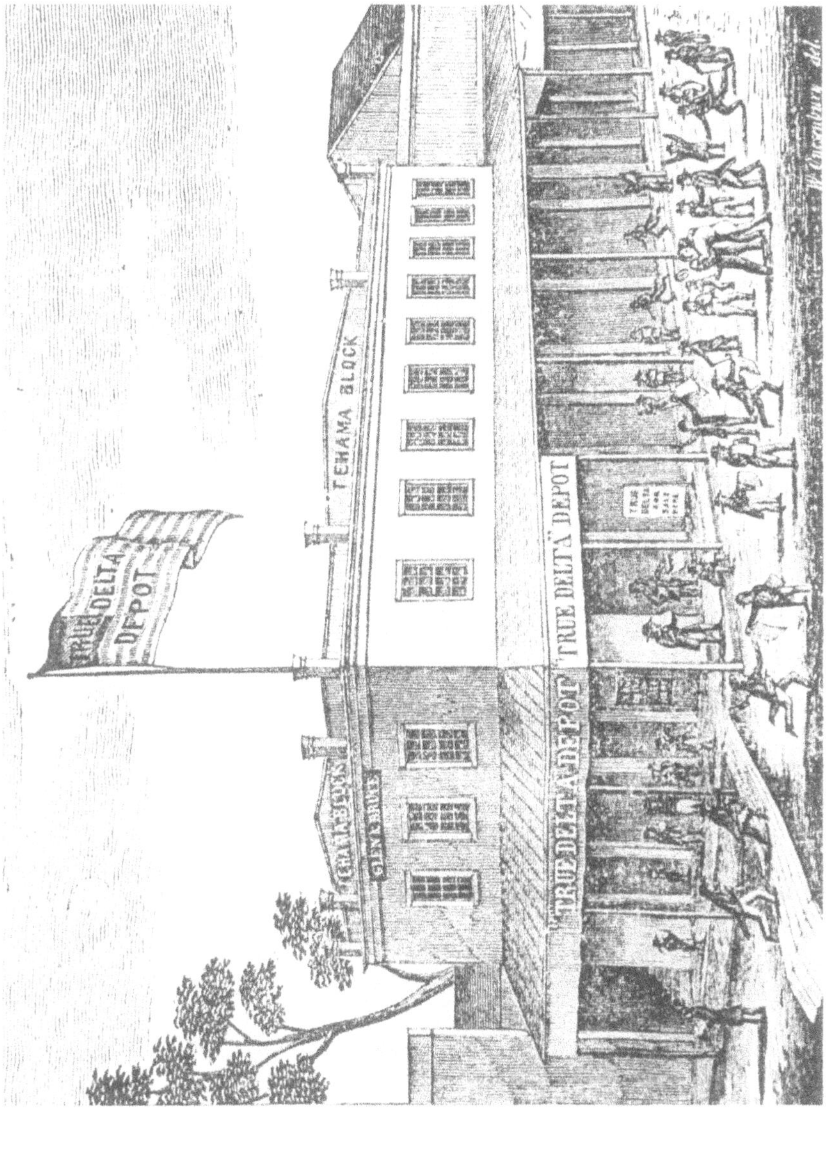
TRUE DELTA DEPOT
TEHAMA BLOCK
TRUE DELTA DEPOT
TRUE DELTA DEPOT

**Figure 4.** The Tehama Block Building was built in 1850 and torn down in 1851, when it was replaced with a brick building.

Courtesy of the Sacramento County Historical Society.

**Figure 5.** The Tehama Block Building in a photograph of 1851. Schliemann's bank may have been located in the building on the extreme left. Courtesy of the Sacramento County Historical Society.

**VIEW IN SACRAMENTO IN THE FIFTIES.**

**Figure 6.** The Tehama Block in 1852 showing the premises of Page, Bacon & Co, Grim & Rumler, G. Frank Smith, E. and R. K. Swift, Adams & Co, Read & Co, and Wells Fargo & Co.

From the *Sacramento Pictorial Union* (January 1853). Courtesy Center for Sacramento History.

My only occupation here being to lend money on mortgages on land and houses, I have nothing to occupy myself with, and, since from my youth I have become accustomed to work from morning till night, I cannot describe the impatience and boredom which torment me.

He concluded this section of the journal by saying that he wanted to leave Sacramento and return to St. Petersburg by way of China.[24]

# 2

# Davidson in San Francisco (1849–52)

Meanwhile Davidson had established himself in San Francisco. His first three letters from there, which were probably sent together, dealt with quicksilver (27/VIII/49), with ways of shipping silver and sending letters,[25] including a request for "a fine pair of Gold Scales" (28/VIII/49), and described the orderly conditions and high prices in San Francisco, and his purchase of "a large Italian wrought Iron Chest" in which to store gold dust (29/VIII/49). On 31 August he wrote that he had bought a house with two rooms for $9,000. "I find that this Country presents a fine field for operations in Exchange and Banking business." He could make capital yield 40 to 50 percent a year, he wrote, "without entering into wild speculations." Money was scarce, however, and he bought small amounts of gold dust for goods or for drafts on London, Paris, Valparaiso, and other places. He asked Rothschild to tell August Belmont in New York to honor his drafts.[26] He also asked them to forward his letters to Paris, because he was "not capable of repeating everything over again." Within two weeks of his arrival, therefore, he had already laid the foundation of his business and begun to assess the economic potential of San Francisco.

On 12 September Davidson wrote a long letter, of which there are two copies: the original sent to Paris and a copy, which differs in many respects, sent to London (see Appendix). In it he discussed many points relating to the production of gold and quicksilver and to the development of San Francisco: the growth of the population, the value of real estate, the scarcity of money (partly because customs duties had to be paid in coin), and the consequent growth in private coinage and the need to establish a mint.[27] He commented in particular on the growing prejudice against Indians (Native Americans), Mexicans, and South Americans, of whom many had left San Francisco, thus reducing the number of both consumers and producers;[28] and he advised the Rothschilds not to send goods for sale in view of the fluctuating demand. On the whole, however, Davidson had an optimistic view of the future of the city. "A very good business may be done here in discounting, and likewise in the purchase and sale of Gold Dust," he wrote, and proposed the establishment of "a sort of bank" to issue notes payable on various places "on receipt of the Gold Dust."

The profits arising from a bank here would be very great, in the first instance you would have the benefit of the difference between the value of one ounce of gold dust and $16 currency, secondly the use of the money for the purpose of discounting, and lastly a decided gain from the number of notes that would be lost, burnt or otherwise destroyed and which consequently could never be redeemed.

Toward the end of the letter he stressed the variety of business opportunities in San Francisco and his confidence that, if the Rothschilds granted him sufficient power, "You will later have no cause to repent having done so" (Figures 7 and 8).

In London, however, the winds were blowing in the opposite direction, and on 16 September 1849, four days after Davidson wrote this letter and long before it and his other letters could have arrived, Rothschild wrote him a long letter instructing him, among other things, to "confine your operations merely to the purchase of gold dust," to buy no land for them in California, and to advance no duties for imported goods.[29] Much of the letter is concerned with affairs in Valparaiso and with general news of the family and business, which had improved since the end of the war in Hungary. It also said that Davidson's brother Lionel had gone to Germany. It may have been Lionel who had first informed the Rothschilds (as later reported in his letter of 12 November) of his advice to his brother Benjamin and thus aroused their fears that Davidson might overextend himself without proper authorization.

Even before this letter was written, however, and long before it reached Davidson, the horse was out of the stable. On 29 September Davidson wrote that "my house has arrived from Valparaiso," presumably dismantled, that he had bought a piece of land for it for $4,000,[30] and that he would either sell or rent it, which would yield an annual profit of 40% of its cost. He also repeated that he had bought a small house as an office and dwelling for $9,000, which he described as "a price which will appear to you very extravagant." In this and other letters he discussed the purchase and shipment of gold and quicksilver, bills of exchange on August Belmont in New York, and general conditions in California. On 14 November, Davidson wrote, "The most perfect order of tranquility reigns in the Town and in the Mines." In a personal letter to Mayer de Rothschild dated 31 December, however, he gave an account of the fire on 24 December and wrote, "However large the profits may be from the Gold Dust transactions, this is a most disgusting place to reside in."[31]

**Figure 7.** View of San Francisco in 1850.

From Theodore T. Johnson, *Sights in the Gold Region and Scenes by the Way*, 2nd ed. (New York, 1850), frontispiece.

THE PRINCIPAL STREET OF SAN FRANCISCO.

**Figure 8.** Principal Street of San Francisco.
From William Redmond Ryan, *Personal Adventures in Upper and Lower California in 1848–9*, 2 vols. (London, 1850), 1, frontispiece.

Davidson's business continued to grow in the following year and a half. In contemporary sources his bank is mentioned as one of the earliest—probably the third, though the precise order varies—and strongest banks in San Francisco.[32] James King of William, writing in 1855, said that "With the exception of Mr. Davidson, who is the agent of Rothschilds, we doubt whether any banking house in this city at the close of the year '49 possessed a cash capital over $100,000."[33] "Davidson, B., ag't Rothschilds, banker" on Commercial Street between Montgomery and Kearny was listed in the 1850 *San Francisco City Directory* .[34] In subsequent directories the address is given as on the northwest corner of Montgomery and Commercial Streets and Davidson's house, in 1856, on the north side of Sutter Street between Powell and Stockton.[35] The building, which he used as a dwelling and office, was apparently destroyed in the fire of 3–4 May 1850, and in his letters written shortly afterward he stressed the need for a fireproof brick building (12/V/50, 17/V/50). On 28 May, he wrote that he had bought some land for $16,500 and that the new building, in spite of the expense, must be in brick. This is presumably the building described in the account of the great fire dated 6 May 1851 and published in the London *Times* of 9 July, of which the writer had an office "in the house of Mr. Davidson, the agent of the Messrs. Rothschild, situated *then* at the north-west corner of Commercial and Montgomery streets."

> This house was so substantially built, and so protected by iron shutters, . . . covered on the top by a layer of Roman cement, and another layer of bricks embedded in cement; all this covered with zinc plates soldered, and the whole guarded by parapets carried up breast-high above the roof . . . that it was considered fireproof.

It was totally destroyed by the fire, however, and replaced by the building shown in Figure 9.

He wrote on 30 June that he had one clerk to whom he paid $4000 a year—only slightly less than the writer of this book earned in his first professional position in 1955—and that he would have some relief from his "constant labour ... now that Mr May has arrived." Johannes May had been a clerk in the Rothschild house in Frankfurt and was apparently sent out to keep an eye on Davidson.[36] He soon became Davidson's partner, and a letter written (in French) on 14 July 1850 is signed "Davidson and May." In two letters dated 14 January 1851, Davidson said that from 1

**Figure 9.** Bank of B. Davidson labeled "Agency for Rothschild." The building (especially the parapet) seems to correspond to the description in the London *Times* article cited in the text.

From an undated photograph among the Davidson Papers.

February he would sign all drafts jointly with May and asked Rothschild to send all new letters of credit "in our joint names." Already on 15 July 1850 Davidson had written (somewhat defensively) that although May "may *write* more than I am in the habit of doing," it was not "through his instrumentality that everything is carried out," because he could neither speak nor write English.[37]

Rothschild provided Davidson with a credit of £40,000, which was renewed monthly and could be drawn on their houses in London, Paris, Frankfurt, Naples, and Vienna or on Belmont in New York and other places.[38] He appears not to have used it, however, and its main purpose, like his real estate holdings, was to guarantee his credit in San Francisco. His principal business was buying gold to send to Rothschild and selling bills of exchange, which provided the money to buy the gold (Figures 10 and 11).[39] He had to sell bills, he wrote on 4 April 1852, in order to buy gold. There are countless references in his letters to the supply and value of gold dust, which at this time varied between $15.50 and $16.50 an ounce, depending on the quality, market, and season of the year.[40] Davidson compared gold digging to gambling (31/I/50) and wrote that the price of gold was determined less by its value in Europe than by local circumstances, including the weather, because rain brought the gold to the surface (4/III/51). "Never was there a place where the uncertainty of human events is so much felt as here" (30/VI/51).

He also discussed quicksilver, for which the demand was low, he wrote, "because it appears that hitherto the Miners have found it more expensive and laborious to employ the amalgamating process than the simple wash- ing, nevertheless quicksilver must ultimately be used in the extraction of Gold" (31/X/50).[41] There are a number of references in both Davidson's and Schliemann's letters to so-called quicksilver gold and quicksilver gold dust, which had been extracted after being amalgamated with quicksilver and which was somewhat less valuable than regular gold dust. "All the gold that we purchase," he wrote on 31 January 1852, came "from the beds of the rivers and from the dry diggings" (Figures 12 and 13). There are also some references to quartz or mined gold, for which Davidson was never enthusiastic. He considered most of the mining companies, with the exception of the Nouveau Monde, to be risky if not fraudulent (31/V/52; 14/VIII/52).[42] Smelted gold and natural nuggets, on the other

**Figure 10.** A "second of exchange" dated 13 January 1853.
From the California Historical Society.

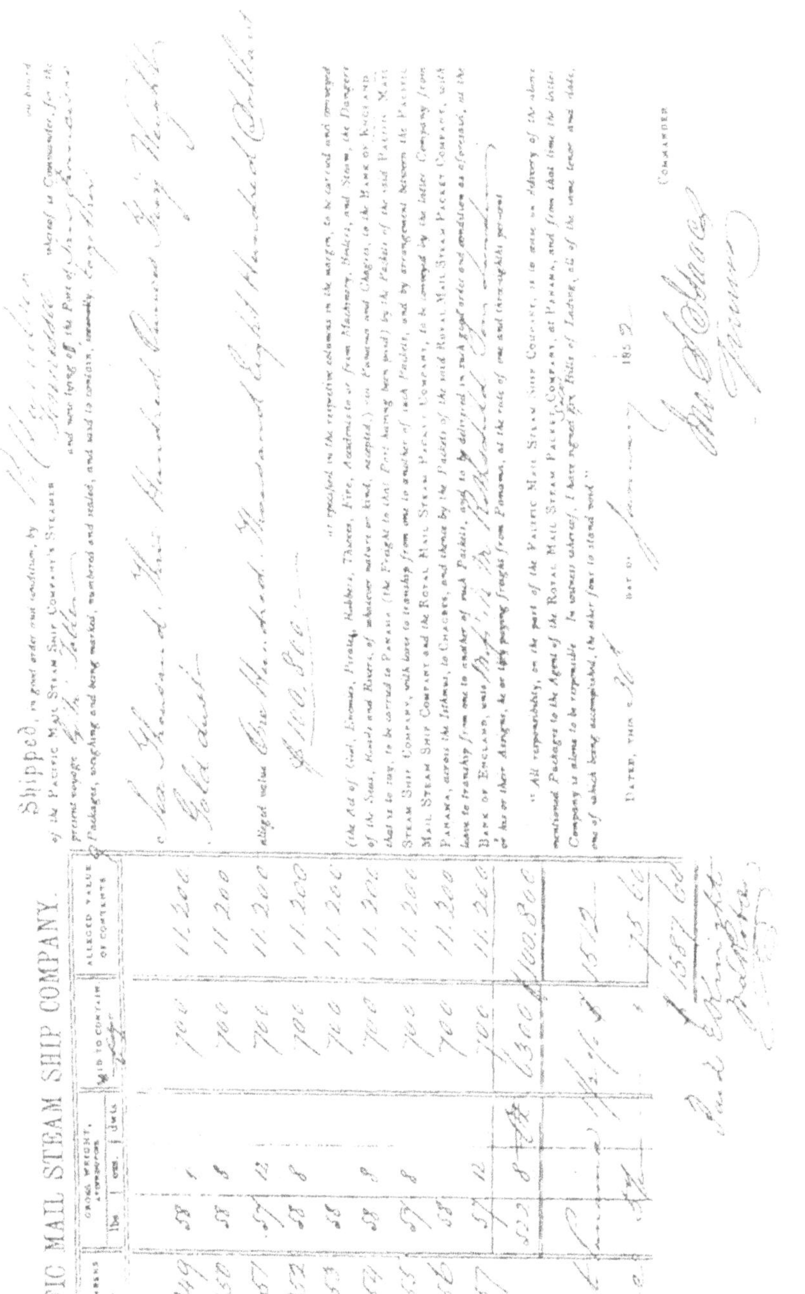

**Figure 11.** Bill of Lading dated 30 January 1852.
From the California Historical Society.

**Figure 12.** The Stanislaus mine.
Reproduced from William Redmond Ryan, *Personal Adventures in Upper and Lower California in 1848–9*, 2 vols. (London, 1850). I.

**Figure 13.** This drawing shows the techniques used by the prospectors from whom Davidson and Schliemann bought gold. From the unpublished sketchbook of Emil Lehman, Center for Sacramento History, reproduced by permission.

hand, were more valuable than dust (22 and 31/I/51). In his letter of 14 January 1851 Davidson described a nugget weighing 159.17 ounces—"one of the largest specimens of pure gold which I have ever seen"—that he bought for $17.20 an ounce (a total of $2750) and hoped to purchase for himself if the Rothschilds did not want it.

Davidson's greatest problem was finding coin with which to buy gold, because many sellers did not want drafts. "At present the only Currency here is the Gold Dust and Coin," he wrote on 31 January 1850, and expressed the need for a regular banking establishment to issue notes.[43] In the same letter he said that he had received $65,000 from Mssrs Davidson in Mexico (that is, his brother Lionel) and had asked them to send $50,000 a month, which was later raised to $100,000 (14/IX/50). The sellers of gold preferred American to Mexican coin, however, and Davidson repeatedly stressed his need for specie. In his letter of 31 March 1851 he mentioned the establishment of a United States assay office that smelted gold, stamped bars of $50 and above, and issued coins of $50 and above for a charge of $2^3/4$ percent, which contributed, he wrote, to the rise in price of gold.[44] The real need was for small coins, and private coiners issued coins of $5, $10, and $20 "in imitation of the U.S. money," which the bankers sought to discourage by paying $17 an ounce for gold dust and charging more for drafts. There are numerous references to these so-called "slugs" in both Davidson's and Schliemann's letters.[45]

Davidson also frequently referred to business rivals and competitors. On 30 April 1850, he wrote that "All the large Mexican houses have Agents here now," and on 30 June 1850 he remarked on the establishment of "a number of new houses . . . who will undoubtedly resort to every measure in order to secure a share of the Business." The competition seems to have been particularly fierce in late 1851 and early 1852. Davidson commented on "the great competition" in his letter of 30 September 1851, in which he said that some houses seemed to do business for nothing "with the sole view of securing the remittance of large sums rather than to allow the same to fall into the hands of other parties." On 14 November he wrote, "There is so much competition and so many engaged in the Gold Dust trade that it is not without the greatest difficulty that one can secure any large parcels," and on 14 December that

The opposition in the Gold Dust trade is growing worse and worse, and all sorts of tricks are resorted to by the Express Companies and the Bankers in order to attract the miners to their respective offices both here and in all the inland towns, where the same price is paid for GoldDust as here.[46]

The largest California banks at this time were Page, Bacon & Co. and Adams & Co., both of which had resources of between $1,000,000 and $2,000,000 and made monthly shipments of gold worth between $400,000 and $800,000.[47] When Wells Fargo was established in San Francisco in 1852 it was said to be in the third rank of bankers, and Davidson in the second rank, with a capital of $1,000,000 at his disposal and monthly shipments of bullion worth between $150,000 and $300,000.[48] Davidson's largest shipment in this period was just over $270,000 (31/VIII/52), and most of his shipments were closer to $100,000 and some as low as $50,000 (Figure 11). These comparatively small amounts were a source of grievance to the Rothschilds. Davidson's standing as a banker was clearly high, however, and in the bank run of February 1855, when both Page, Bacon and Adams failed, Davidson and Drexel, Sather and Church were said to have "met the attack with proper resistance and survived."[49] According to William Tecumseh Sherman, in a letter dated 25 February 1855, Davidson paid out $800,000 in eight days, and "Davidson is the recipient of the aid of the Jews and foreigners. He stood the run well and from foreigners raised outside assistance."[50]

In addition to buying and shipping gold and selling drafts, Davidson engaged in various other business activities, especially real estate. He wrote in his letter of 12 September 1849, shortly after his arrival in San Francisco, that, "The largest fortunes that have been gained here have been accumulated by the speculators in building lots and houses, which sell and rent at perfectly fabulous rates" and added later that "For the next 6 months the construction of houses will be a most profitable business, as one years rent at present rates would suffice to pay the whole cost of the construction and of the land." Davidson's later letters show that the value of land and houses in fact fluctuated owing both to the changes in demand and to the prevalence of fires, but he nevertheless, and in spite of the Rothschilds' explicit instructions not to acquire land, invested heavily in real estate, partly in order to support the credit of the bank. His 1851–52 tax bill shows that he owned lots on Clay, Commercial, Montgomery, California, Beach, and Water Streets, worth a total of

$59,000, and had personal property worth $100,000.[51] Not all of these turned out as well as he hoped. He referred in a letter of 31 August 1851 to depressed business conditions and the reduced rents from the Clay Street property. He suffered both materially and personally in the great fire of May 1851, which destroyed his bank and from which he escaped, according to the account in the *Times* cited previously, "by forcing his way from the roof, through the fire, in wet clothes with a damp sponge in his mouth, and slipping down the walls by means of a blanket into Montgomery-street." He rebuilt the bank, however, and extended his real estate holdings not only in San Francisco but also in Sacramento, which will be discussed later; he may also have had property in San Diego.[52]

Davidson also engaged in miscellaneous commercial activities. As early as his letter of 30 August 1849, he referred to buying gold dust for goods, showing that he kept a kind of shop. He seems to have bought cigars for the Rothschilds when he was in South America, and he commented on the demand in San Francisco for sugar and "segars" (30/IV/50). There are many references to tobacco in other letters and in the American Letter Book in the Rothschild Archives. He took advantage of the growing trade with the Far East (19/IV/50 and 31/V/50), and in October 1850 he entrusted drafts for £6000 to Mr. Charles Hitchcock of the *Sarah Warren* for trade in Manila and China (25/X/50, 31/X/50). A list dated 14 February 1851 shows that in return he received merchandise (including sugar, currants, coffee, cigars, and spices) valued at almost $20,000, which he eventually disposed of at the comparatively small profit of $3622.25. A "Barley speculation," however, yielded a profit of $15,000 on an outlay of $18,000 (30/VI/50). He also speculated in lumber and brick (30/V/51), which were at a premium after the Great Fire, and he even had "an assortment of . . . pianos for sale" according to a letter from Schliemann (3/XII), who wanted to buy one for a friend. In a letter dated 25 April 1851, the vice-consul of France described a large shipment of coal, wine, liquor, cognac, vinegar, glassware, nails, and diverse merchandise to Mr. Davidson "agent of M. M. de Rothschild at San Francisco,"[53] and in his letters of 31 July and 14 August 1851 Davidson referred to selling French wine and to the demand for French wines and spirits.

It is uncertain how much of this business was on behalf of Rothschild and how much on his own account, but it is clear that the Rothschilds both in London and Paris were nervous that Davidson was overextending himself, or, more likely, themselves, because most of the money was theirs. They apparently accused him of disregarding their instructions, because on 31 January 1850 he wrote that he was taking no salary because he had acted in part against their instructions.[54] At the same time, they regarded him as insufficiently energetic, perhaps with regard to sending gold, and wrote (according to Davidson's letter of 28 April 1850) to say that they were dissatisfied with his "inactivity and want of energy" and planned to send another agent to act in his stead. This may have been Johannes May, who arrived in San Francisco soon afterward. The suspicions of the Rothschilds were also inspired by reports brought back by visitors. In his letter of 30 May 1850, Davidson sought "to confront every accusation and every complaint which you have uttered" and asked them "to take into consideration the great anxiety and mental excitement under which I have been labouring since the late calamity," presumably the fire of 3–4 May.[55] The three principal accusations were (1) his expenditures on land and houses; (2) that he was extravagant and played "the part of a 'grand man'," which he strenuously denied, stressing the high cost of living in San Francisco;[56] and (3) his speculations in lumber and brick, which he said were on his own account.[57]

The most serious matter was real estate, with regard to which Davidson assured the Rothschilds that the investments were profitable, that he needed a brick house, and that "although the land you own stands in my name" he would draw up a statement that it belongs to them, even if this affects his credit. Seventeen months later, on 31 October 1851, he wrote that he had drawn up an agreement "in accordance with your instructions" showing "that the Real Estate which I purchased with your funds belongs to yourselves, and that I only hold the same with the improvements thereon as your Agent and subject to your control." No copy of this agreement is known to exist in the Rothschild Archives. Among the Davidson papers, however, there is a draft, drawn up in San Francisco and dated 27 October 1851, with many alterations, of an agreement between Benjamin Davidson "of the first part" and Lionel, Anthony, and Meyer [sic] de Rothschild "of the second part." It describes the properties on Clay Street (with "½ interest" written in the margin, suggesting that Davidson owned the other half), Commercial and Montgomery Streets (marked "The Banking House property" in the margin), and Battery Street ("the Water lot") "and the corrugated Iron house now standing thereon."[58] They were in Davidson's name "the more effectually to enable him to conduct the Banking and other business of his said Employers in San Francisco," but they were bought with Rothschild money and held by Davidson "merely as Agent aforesaid and for the use and benefit of his said Employers," and he will convey title to them whenever they wish. It is unknown whether this agreement was ever finalized. On 14 October 1851 Davidson wrote, as if in response to instructions from the Rothschilds, that "I am to keep you constantly informed of the value of your land and not to dispose of any part of it until I receive instructions from you to that effect," to which he objected that the value of land (like gold dust) fluctuated and that it would take four months to inform them and hear back.

The tension between Davidson and the Rothschilds was exacerbated by the rivalry between their London and Paris houses, each of which wanted to profit from the gold trade. According to a letter from Lionel Davidson to Mayer de Rothschild dated 22 November 1849, Baron James in Paris was infuriated "to an extraordinary pitch" by a letter from his nephew Mayer in London saying that he had instructed Davidson "*not to send any gold to France.*"[59] No less important was the fact that in spite of their concerns about Davidson the Rothschilds were instinctive entrepreneurs and could hardly resist the opportunities to make a profit presented by San Francisco, in spite of Davidson's warnings of the risks.[60] Davidson's letter of 28 May 1850 shows that they were keen to expand their business in San Francisco. Inspired no doubt by reports of the demand for tobacco, they sent him some "Segars" from London for sale in San Francisco (30/XI/50), and the French ship laden with merchandise mentioned previously was probably sent by the Paris house. In spite of their instructions to Davidson to restrict himself to the gold trade, the Rothschilds expected him to engage in other types of business, which sometimes created difficulties.

The most serious and long-lasting trouble was caused by some iron houses, the first reference to which was in a letter dated 12 February 1850 from Lionel Davidson to the Rothschilds acknowledging receipt of up to £11,000 "in reimbursement of an order for certain Iron Houses to be shipped to Califomia."[61] Iron was widely used as a building material in the first half of the nineteenth century owing to its strength and resistance to fire and shock.[62] In March 1850 iron houses were said to be "very abundant" in San Francisco "but in slight demand,"[63] because they stood up to fire less well than expected. The correspondent for the London *Times* wrote on 24 January 1851 that, "The losses upon iron-houses [imported from England] alone have been very great. Houses of iron are totally unsuitable to this climate."[64] It is therefore not surprising that when Davidson heard from his brother about the iron houses he wrote to the Rothschilds, on 14 April 1851, that they would be impossible to sell in the existing market. The final blow was dealt by the Great Fire in May 1851, which according to the London *Times* correspondent writing on 5 May (published 9 July) "settled their value as a protection against fire for ever. Not one of them stood the attack of the fire. . . . They helped the fire materially by the intense heat which they emitted."[65] Davidson referred to the iron houses in many other letters, including one dated 4 December 1851, in which he said that they had given him "more trouble than any affair I have ever had here," and another on 14 April 1852 stressing that "*I* as the Consignee cannot respond for the result of your shipment."[66] It appears that in all he disposed of only three houses in addition to the one built on the Battery Street lot, referred to in the draft agreement of 31 October 1851. John Luck, in his letter of 26 April 1852 (see Appendix), said that the remaining houses were "only worth the price of old iron." It was a serious error on the part of Rothschild to send them in the first place and one that damaged the relations with Davidson.

Davidson's letters are, finally, a valuable source of information on the early history of California generally, shedding light not only on the growth in population—"The emigration continues to be as great as ever," he wrote on 19 April 1850[67]—but also on the departure of some of the population, including the Mexicans and South Americans (who have already been mentioned) and the Europeans who left owing to disappointment in the gold fields, especially after the discovery of gold in Australia (14/VIII/51, 31/VIII/51). Under Spanish law, which was still in force when Davidson first came to California, before it became a state, any mine that was not worked for three months could be claimed (29/IX/49, 10/X/49). The subsequent changes were not all for the better. In spite of his original impression of order and tranquility, Davidson soon objected to "the Class of Men who are at the head of the Government" (31/III/50), to the corruption of the government and its bad administration (15/VII/50), and to the "maladministration" of the state and the city (31/I/52). Citizens sometimes took the law into their own hands. When "a serious disturbance" developed between the squatters and property owners in Sacramento, the merchants organized into a private company called the California Guards (15/VIII/50); "some slight local disturbances" in San Francisco, probably in connection with the Great Fire, led to the imposition of Lynch law (4/III/51); and after a fire that may have been started by an arsonist, a Secret Committee was appointed to act "as did the inquisition in former ages" (30/VI/51).[68] The social, economic, and political situation in California was in constant flux and Davidson had to struggle to stay on his feet.

# 3

## Schliemann in Sacramento (1851–52)

There is no evidence that Davidson had any contact with Schliemann before September 1851. In the Notes of Drafts sent with his letter to Rothschild on 14 September Davidson listed fifteen drafts of various amounts, totaling £225, for Henry Schliemann, and on the invoice dated 30 September he recorded that Rothschild would be debited "with the amount of your charge on the protested bill of Schliemann on Amsterdam," about which nothing more is known. In a letter dated 30 September he wrote, "We now further beg you will take note that Mr Henry Schliemann of Sacramento City has valued upon yourselves" three drafts for $494.53 "which amount we have received and placed to your credit and therefore request that you will protect the above draft on presentation." Pasted onto the bottom of this letter is a slip with Schliemann's characteristic signature cut out from a letter. The Notes of Drafts sent with this letter include nineteen drafts in varying amounts totaling £1570. Fourteen more drafts totaling £860 were listed on 14 October; four totaling £1000 on 31 October; ten totaling £330 on 15 January 1852; two batches, one of thirteen drafts totaling £340 and the other of eight drafts (for H. Schliemann & Co.) totaling £170 on 31 January; eighteen drafts (for H. Schliemann & Co.) totaling £340 on 17 February; fifteen (for H. Schliemann & Co.) totaling £660 on 1 March; and one for £900 on 17 April.[69] All of these drafts are "at sight" and at rates varying between 46% and 47.5%. They show that between September 1851 and April 1852 Schliemann was regularly buying drafts from Davidson and that together they added up to a considerable sum, perhaps as much as £7000.

Davidson appointed Schliemann his agent in Sacramento probably early in September 1851, which corresponds with the statement in his letter of 17 April 1852[70] that Schliemann had been his agent "for the purchase of Gold dust and Sale of Bills" for the past seven months, with May's letter of the same date, and also with the entries in Schliemann's journal (making some allowance for ante-dating) that he established a bank to buy gold dust and sell "exchange" (June 1851) and that "My purchases go for the most part to the house of Rothschild in London," of which the branch in San Francisco supplied him with coin (31 July/1 September).[71] It is unknown how much of Davidson's total shipments of gold came from Schliemann, but he was certainly a major supplier. In the advertisement dated 10 September for the "Banking House of Henry Schliemann" in the Sacramento *Daily Union* he offered to buy 3000 ounces of gold dust for $17 an ounce in gold coin or "for Drafts At Par on San Francisco or for Drafts of B. Davidson House of Rothschild at San Francisco on the U.S. and Europe."[72] This was changed in subsequent advertisements (which were still dated 10 September) to describe the "Banking House of Henry Schliemann" as "Agent of B. Davidson, House of Rothschild, in San Francisco" and to specify the places of exchange as the Rothschild houses and correspondents in New York, New Orleans, London, Paris, Hamburg, and Frankfurt. Copies of this advertisement (always dated 10 September) appeared in the *Daily Union* and in other newspapers from time to time until 6 April 1852. Schliemann referred in his letter of 20 December 1851 to "our continual advertising in all papers."[73]

These advertisements show that the relation between Davidson and Schliemann was open and recognized. Though no formal agreement between them is known, evidence that an agreement existed is found in several of Schliemann's letters. Writing to Davidson on 17 (?) October he described his appointment for one year

> as your agent for Sacramento City under the following conditions. I have to deposit with you a Bond of Security the sum of Twenty Thousand Dollars . . . ,[74] you furnish me with sufficient Coin for the purchase of Golddust, of which I have to dispatch to you daily, Sundays excepted, all what I can possibly obtain in good clean quality fine [?free] from quartz and stone. . . . During the above term I cannot connect myself with any other house in your city and should I without your knowledge send Golddust either on my own or on foreign account to the Assay-office or to any banker of your city, then my Bond is forfeited and our connection broken forever.

He went on to specify that the rates for Davidson's drafts were fixed for the first two months at 47½ on London and 1½ percent premium on New York—"if I can sell better it is my advantage" —, that he is to receive ½ percent commission on clean gold dust bought at $17 per ounce, and that he will divide with Davidson the return-commission from the assayer Moffat on parcels of quicksilver dust and gold "which I send you on foreign account for Coinage."[75] Finally, he said that his office and clerks, of whom he was to keep two,[76] were at his expense and that he had to "keep my books constantly open for your or Mr May's inspection."

Two days later (though the date, as with the previous letter, is uncertain) Schliemann wrote:

In consequence of your desire to enter as a private partner in my agency for your goodself, which by your letter of yesterday you conferred upon me for the term of one year from date, I sign from this day of: Hy Schliemann & Co. of which please take note. I beg to repeat to you, that the *utmost care* and *most scrupulous attention* shall constantly be paid to the promotion of your interest.[77]

A further specification, arising out of the reference in the agreement to the "good clean quality" of the gold dust, was spelled out in Schliemann's letter of 19 December (written in German) saying that Davidson did not have to accept unsatisfactory dust.[78] With minor adjustments (principally in the permitted cost of gold, the amount of gold to be shipped each week, and the sale of drafts), this agreement remained in force for the duration of their partnership.

Schliemann was a connoisseur of gold dust, or Golddust (and occasionally Goldust), as he called it, and of the "Golddusttrade." He described the gold in different letters as "beautiful," "splendid," "purest," "perfectly clean," "very fine," and "superior" or, at the opposite end of the scale of quality, as "coarse" and "unclean." He distinguished "the best scaly Feather river gold" from "scaly Yuba-river gold" and "Nevada quartz gold," which he considered inferior to gold dust. He asked Davidson several times to send him "good seves [sieves] *just exactly like yours*" (9/ XI) to sift the dust and extract sand and stones. He also encountered bogus gold, which was much in circulation and of which a Mexican brought in a large lot one day (24/XII; 16/II).

Schliemann acquired gold both directly by purchase from prospectors (see Figures 12 and 13) and through a network of traveling and resident agents who bought gold for him. On 16 November he wrote:

We should hardly be able to send you 150 ounces per day at present had we not travelling agents and traders who buy for us, all over the country and to whom we pay $17 with 1/2% commission—from Mr Whetstone in Shasta particularly we get large lots.

Whetstone (Whitestone or Whitstone, as it is sometimes spelled) first appeared in a letter of 26 October, in which Schliemann wrote "that one of our depositors a certain Henry S. Whitstone has left yesterday for Shasta to establish there for us an agency for the purchase of Golddust and the sale of your drafts."[79] On 20 December Schliemann reported to

Davidson that he "had a long conversation with Mr Whetstone regarding the Golddust business in Shasta" and that he was

> convinced that an extremely profitable business might be done with that place if we manage it properly, the price being there only $10 per ounce for dust such as you get by this conveyance. All we have to do is to join Mr Whetstone, another confidential trustworthy person, who superintends the business and takes care of the sale of Exchange.

Whetstone actually went to San Francisco to discuss the matter with Davidson (21/XII), who rejected the idea, and Whetstone "now intends to continue buying and shipping to us as heretofore, viz. on his own account and at his own risk" (24/XII).

Schliemann's agent in Nevada City was Lloyd Tevis, who later became president of Wells Fargo and spoke admiring words (cited on p. 59) about Schliemann at the time of his death. He "has about $20,000 capital of his own and all the risk of the money up and down is of course on his account," according to Schliemann (28/X, replying to Davidson's letter of 27/X), who also referred to Tevis in his letters of 2 and 9 November.[80] In other letters Schliemann mentioned "our traveling agent Steadly" (6/XI), "a man who buys for us in Hangtown" (11/XI), and Harris Levingstein as one of "several other Shasta traders" who, in addition to Whetstone, were "in the habit of bringing their Dust to us which is immediately sent to you" (19/XI). Schliemann was concerned that Davidson might cut him out by dealing directly with these agents, and in the letter of 28 October Schliemann assured Davidson that he ran no risk from "our agencies in the interior" and asked him to

> confirm to us in answer to the present, that you will *never and at no time* connect direct with any of our agents and particularly not with Lloyd Tevis and that you will always refer him to us whenever he addresses you regarding exchange or Gold dust business.

Davidson apparently gave him the desired confirmation, as Schliemann reminded him on 9 November that he had given "the surest promise *not* to operate directly with Mr Tevis and to let everything go through our hands as if a branch-House of ours [for yours?]." It is not known whether Schliemann was concerned only about his commissions or had something else to hide, such as the exact quantity and cost of the gold he was sending.

Schliemann preferred to obtain gold directly, across the counter, or through his agents rather than from traders, who presumably charged a commission. "Today it was again *exceedingly Dull* in the Gold trade," he wrote on 10 November, "and almost no miners coming in we were compelled to purchase exclusively from Traders who are very stubborn to deal with." On 20 November he wrote:

> To make up a little shipment for you we were obliged to buy some parcels from our City Traders at $16^1/4$ to $16^1/2$ which when cleaned stood us at $17^1/4 and thus, although we lose money on our shipment of this day [because Davidson paid only $17] yet we are sorry to say that it is not so perfectly clean as you are accustomed to get from us.

He complained that gold dust was scarce and "miners and traders more and more stubborn" on 11 February, when he also wrote:

> We have for the last two months been anticipating from week to week an amelioration in the Goldtrade, but the lucky period so long desired for, seems to lye still far off. In this strange community we cannot force anything, we leave matters and things to take their own course and the price of Golddust is bound to regulate itself according to the quantity offered and in spite of the most capricious competition the value will come down again to its former standard of $17 as soon as the supply in the market proves to be inadequate to the demand.

Still on 24 March he wrote that he was buying almost exclusively from traders because the miners were not coming into town. Occasionally he bought from bankers (29/XI), but they sold on condition that the gold be put in the mint and that they get the profit, aside from a $^1/2$ percent commission (12/XII).

Schliemann had the same problem Davidson had in procuring coin money, especially in small denominations, with which to buy gold. Sellers wanted coins, though some (but not all) of them would take partial payment in bars and drafts and occasionally slugs and private coinages. In his advertisements Schliemann promised payment in gold coin or drafts, and sometimes in coin only, and he was worried about being called a "humbug" if he ran out of cash (6/XI) and could not fulfill his promises. He was caught between the Scylla of having too little coin and the Charybdis of having too much, which might attract thieves (2/XI). In principle Davidson sent $3000 a day, but this was not always in coin, which he also had difficulty in obtaining, and Schliemann had to buy coin at a premium. He complained on 25 October that "As we did not get any more american Coin from you, we were compelled to go begging from store to store to get some," and on 9 November that he had to pay a premium of $1^{1}/_{2}$ – 2% for coin, for which the demand is "so extensive that we are obliged to have 10 to 12,000 of it continually on hand to supply our customers who have a perfect aversion to ingots."[81] On 6 November he asked for $8000 per diem in bars (raised to $10,000 on 10 November) in addition to $3000 in American coin and $1000 in silver.[82] *"If we have no amer. coin,"* he wrote in his letter of 9 November, underlining the whole sentence, *"we can buy no gold for you."*

Each day, after Schliemann had gathered some gold as best he could, either in gold dust of varying qualities, "smelted gold" from quartz, or quicksilver gold ("Queckes Geld" in German), he had the problems of shipping it to San Francisco, where it had to be assayed. He usually used Gregory Express, which charged $^{1}/_{4}$ percent (19/III) and to which there were many references in his letters.[83] He became increasingly dissatisfied with Gregory Express, however, especially during his illness in January, after which he complained of its "downright impoliteness and inattention" in forwarding letters (17/I). Already, on 20 December, he had written that Adams & Co.,[84] which shipped gold for Page, Bacon & Co. for $200 a month "whatever may be the amount forwarded," were ready to "take on themselves *all* risks of acts of God and robberies or neglect" for $^{1}/_{8}$ percent. On 7 January he urged Davidson "to authorize us to ship through Adams," and when Davidson did not do so, on 17 March Schliemann wrote that he would use Gregory "in majority of cases" but sometimes use Adams. On 19 March, Schliemann wrote that he was under obligation to Adams for having supplied him with small gold coin "without which we can*not* buy any Golddust" and had therefore made "from time to time a little shipment through them to you. Still since you *disapprove* of this system, it shall be broken off on the very beginning and whatever Golddust we may have to ship on our own account goes as heretofore so from this day forward through Gregory" (Figures 14 and 15).[85]

Davidson and Schliemann had similar difficulties with assaying, which they normally entrusted to Moffat & Co. rather than the U.S. assay office.[86] Moffat made an error, however, in stating the fineness of two identical parcels of gold, according to Schliemann (22/X), who wrote on 24 December that a particular assay gave "a new and most striking proof that

**Figure 14.** Cunningham's Wharf, San Francisco, and the Sacramento boats, including the *Senator* and the *New World*. Courtesy of the Center for Sacramento History.

**Figure 15.** The *New World* about 1850.
Courtesy of the Center for Sacramento History.

44

immense frauds are committed in the shaving-shop of Moffat." On 11 February he asked Davidson "to go *yourself* to Moffat and to recommend the above lot to his particular care and attention" because

> We cannot but look with great distrust and diffidence on the negligent or irregular management of Moffat's establishment and not without trembling we look forward to your information regarding the yield of our property entrusted to his hands.

Four days later, on 15 February, he advised Davidson to "sell your surplus Gold on the spot, enjoying a good profit, instead of entrusting it to the insecure hands and irregular arrangements of the Assayer, or instead of ruining your agents' business by restricting their purchases."

Some interesting light on Schliemann's competitors and business practices is thrown by three letters of 16, 22, and 25 November.[87] In the first, after commenting on the growing competition and number of new banks, he wrote:

> The old business-spoilers Adams & Co. have a few days ago established here a Bank close to the Crescent City Hôtel and they will no doubt entirely destroy and ruin the Golddust business, which was already before as *unremunerating and dead as possible*. Mr Page [of Page, Bacon and Co.] and ourselves we had some days ago the idea to put the price up to $17½ for a time in order to crush all the little Bankers and humbuggers and to take advantage after having reached our aim;—but since Adams has established himself in our midst there is no hope of success for us, because when we pay $17½ he will allow $18 and rather ruin himself than stand back.

Six days later, in the second letter, he wrote that gold prices were up to 17¼ to 17½ and that both Page, Bacon & Adams were ready to pay $18 because "they are compelled to remit One Million of Dollars Gold Dust by next Steamer."

> As the prices of 17¼ and 17½ are calculated to break here all the small Bankers and humbuggers who operate with a small capital, so we are convinced that as soon as this most desirable aim is accomplished, the price of the article will recede to its former standard and the Gold Dust trade will then assume a healthy remunerating character and tend its influence to create a general beneficial effect.

In the third letter, written three days later, he reported that he and Page, Bacon had put their price down to $17 but that "in the upper part of the city they continue to pay 17¼," presumably "in order to try our capacity and strength and we are happy to say we have shown them that we do not suffered [sic] ourselves to be trifled with and that we can stand the competition as well as any of them."[88]

In addition to the "Golddusttrade," Schliemann conducted other business for Davidson, in particular the sale of drafts, of which the terms were established in their agreement. In his letters of 23 October and 7 November, Schliemann specified that he could return any drafts he was unable to sell. Davidson had sent some drafts on Belmont, he wrote, "under the express condition and distinct understanding that you take back and give us credit for all what we are not able to dispose of." On 21 December, indeed, he returned "all what we hold of your London [drafts]," which he could not sell. On 12 February he asked for "one more draft" on Paris, and on 20 February he said that he sold each month between $16,500 and $25,000 worth of Davidson's drafts on New York, nine-tenths of which were paid for in $50 slugs (which he probably did not want). In the same letter Schliemann complained that whereas his competitors charged 3–3½ percent for drafts of over $400, Davidson charged 4 percent "at which rate we also sell them," leaving him no commission. For this reason "we must charge a little on bills on London," to which Davidson had objected in some "unjust and very unpleasant remarks."

Schliemann also transacted private business for Davidson, including the purchase and sale of state and city bonds,[89] war bonds, and above all real estate, to which there are many references in Schliemann's letters. Davidson apparently owned eight lots in Sacramento, which Schliemann managed in return for a 5 percent commission, renting and occasionally selling them and paying the income into a separate real estate account.[90] This real estate gave Schliemann a lot of trouble. In November 1851 the tax collector was dissatisfied with the payments on Davidson's property but admitted his error two days later (17 and 19/XI). On 1 March Schliemann wrote that, "We have arrested a Mexican for attempting to set on *fire* the House on lot no. 3" and who was to be tried for arson that day. On 2 March he wrote that the tenant of lot 3, named Bennett, was in financial difficulties and unable to buy it and two days later that Bennett was to "remove" the house, which he offered to sell for $500. Schliemann refused and said he would sell the lot when it was free. During the flood in March 1852 at least one of Davidson's houses was under water, two more unrented, and Schliemann allowed a small reduction in rent on the others.[91]

Schliemann stressed his hard work and attention to duty in both his journal and the letters. On 2 November, when Schliemann's clerk Satrustegui was ill, for instance, he wrote:

> I had a very hard time here during the last week and never a negro slave worked harder than I did. But that is all nothing to the danger of sleeping the night alone with so immense amounts of Gold in Cash. I always spent the nights in a feverish horror and loaded pistols in both hands, the noise of a mouse or rat struck me with terror. . . . In one word: it was a most awful time, but now thanks to Heaven my sufferings are nearly passed, for the clerk I engaged here at $250 per month comes tomorrow.[92]

His health was also a concern, especially in January 1852, when he had to leave Sacramento for a few days in the healthier climate of San Jose.[93] Satrustegui wrote him on 13 January, hoping that he would "speedily recover of your indisposition," and kept him informed of business affairs. When he came back to Sacramento he drew up his will and sent it to Davidson on the understanding that it would be returned to him when he left California and with instructions that in the event of his death his property in Rothschild bills on London should he sent to his sister Louise in Rostock (17/I; 20/I), who was also mentioned in his letter of 10 November.

With regard to his finances, he always distinguished Davidson's ("your account") from his own account, which he sometimes called "nostro conto" in Italian and described as "for our account and at our risk."[94] He also kept a separate real estate account and a "foreign account (received by us)," which referred to accounts for assets entrusted to him by others (8/XI). The $20,000 Bond of Security was likewise kept in a separate account by Davidson. None of the records of these accounts are known to have survived. The number of drafts bought from Davidson by Schliemann show that he built up a substantial fortune during the period they worked together, in spite of his frequent complaints that he was making nothing. On 7 November, for instance, he wrote:

> At present we have not the least benefit from your business, and very likely we shall not have any before the rainy season sets in and brings down the

large piles of Golddust accumulated in the miners' hands during the summer. What little commission we have now from you is doubly taken away and swallowed up by the premium we pay here on american Gold.

In his German letter of 19 December complaining of the cost of gold he said that "Wir selbst machen jetzt nicht 1/10 unseren ungeheuren Kosten."

Schliemann supplemented his income from his dealings with Davidson by other business, especially the "foreign" accounts, about which little is known. Whether or not they were in breach of his contract with Davidson is unclear, but he made no secret of them in his letters to Davidson. Most problematical are his dealings with Baring Bros., to whom fourteen letters in his letter-book were addressed and to whom he certainly shipped gold, though apparently not in San Francisco, which was specifically forbidden by the agreement. In his letter of 26 December he described "a box containing One Thousand Ounzes Golddust, which you will be so kind as to ship by steamer of 1st January to Messrs Baring Bros and Co of London" and added, after spelling out some details of the shipping and insurance, "Excuse for the trouble we cause you by our shipments; but please be assured of our constant greatest desire to serve you and further your interests where and whenever we possibly can."[95]

Schliemann's letter of 28 December, which is marked "Private," shows that Davidson objected not to his shipping gold to Baring but to "the impropriety of my depositing money with B. B. and Co." To this Schliemann replied that in view of his old friendship with Baring Brothers, which went back to his time in St. Petersburg, he thought it only fair to divide his means and to deposit half with Baring and to leave the rest in California as a bond of security with Davidson. In any case, he said, he thought Rothschild business was "on an immense scale" and that they would not take deposits. In the future, however, he would draw on Baring and remit his drafts to Rothschild. He did not stop sending gold to Baring, however, and referred in his letter of 1 March to sending them five hundred ounces of the "Finest California Gold Dust."

Davidson seems to have had no objection to Schliemann's shipping gold (presumably in excess of the stipulated amount for himself) to Europe so long as he did not send it to other banks in San Francisco. In a letter dated 21 March Schliemann strongly denied that he was

in the habit either of selling the flower and bulk of our purchases here or of shipping the same to one of your competitors in San Francisco, which accusation we reject with scorn and indignation, as a foul, mean and absurd calumny of desperate foes;—never one ounze of Gold has been sold here on the spot by us ever since this Bank was opened, never one pennyweight was shipped to other San Francisco parties.

This was only one of many sources of disagreement between Davidson and Schliemann, who frequently complained of Davidson's treatment of him, especially his failure to send coin and drafts. "But, Dear Sir, why do you treat me so 'step-motherly'," he wrote on 21 October. "Why do you send me so little Coin?" And on 2 November: "But why, dear Sir, for the sake of Heaven do you not send me an assortment of drafts in sums of $50, 75, 100, 150, 200, 300, 400 and 500, as you are aware that I am bare . . . and you can easily enough conceive, that when my agents send down Dust for *Rothschild's drafts* I cannot be so base as to send them poor Schliemann's bills, which latter have no value before they are paid?" On 6 November he again expressed "our *utmost astonishment* at your manner of treating us" by not sending drafts.

Davidson was clearly sometimes short-tempered with Schliemann, who wrote on 24 October that "We refrain from answering your private letter, the same having been written in an hour of great excitement." Davidson seems to have been upset by Schliemann's refusal to take a friend named Richter as a partner and also by Schliemann's request to find him an additional clerk, which put Davidson into "an angry passion." The following day Schliemann protested "most decidedly" the injustice of Davidson's keeping some quicksilver dust "with ½ per cent commission for your account." On 20 February Schliemann answered Davidson's "unjust and very unpleasant remarks regarding the rate of Exchange charged by us for your drafts on London," for which he charged extra in order to make up for his lack of profit on the drafts on New York. There was constant bickering over the return commission and other relatively small sums of money, and over short weight, "those trifling differences," as Schliemann called them, to which he asked Davidson not to return (8/XI) and which he attributed among other things to reusing the bags in which the gold was sent (20/III). The principal disagreements were over the quality of the gold dust, which, as has been seen, Davidson sometimes refused to accept.[96] This annoyed Schliemann, who wrote on 12 December

that we have grown *completely sick* of your continual complaints as to the quality of our Golddust shipments, and in order to exclude henceforward

from our correspondence this highly disagreeable and unpleasant topic . . .
*we herewith request you to bring at a dead loss to our debit any quantity of*
*Sand which you may be able to extricate from our shipments at 17$^1$/4.*—But
please do so in a few words only and not with complaints, which we deserve
in *no* way . . . . Since we mix in the parcels inferior Gold all the ingredients
which we pick out from between the fine lots, the former may perhaps some
times not be very welcome purchases to you, but we repeat to you herewith
what we agreed with you in October and with Mr May a few days ago, viz.
that you have at all times the privelege [sic] to refuse our shipments [of]
inferior Dust and to put them on our account and at our risk in the Assay-
office, crediting us the returns.

# 4

## Parting of the Ways

These excerpts from their correspondence show that from the beginning the relations between Davidson and Schliemann were uneasy. There is little evidence in Schliemann's letters of any personal warmth between them. They were business partners rather than friends and always regarded each other with a certain measure of distrust. Some scholars have attributed the breakdown of the partnership and Schliemann's departure from Sacramento to Davidson's dissatisfaction, especially with Schliemann's choice of carrier and short weight, and his suspicion that Schliemann was sending gold to other bankers in San Francisco.[97] According to the Sacramento *Daily Union* (8 April), Schliemann left town to attend to family matters. Others have attributed Schliemann's departure to his dislike of Sacramento, which is vigorously expressed in both his journal and his letters, though in a letter of 20 February he wrote that "We have the greatest confidence in the hourly more and more progressing, flourishing conditions of our City."[98] In particular he was worried about his health, and he wrote in the journal that after catching the fever again on 17 March he decided to leave and to "give over my business to Mr. B. Davidson, agent for Rothschild in San Francisco and return as soon as possible to my beloved Russia, for I feel I should not survive if I caught another time the fever."[99] Because Schliemann's letters show that he was at work in Sacramento from March 17 to April 6, "This third attack of fever is a sheer fabrication," according to Traill, who argued that "It seems designed to provide a respectable explanation for the sudden closure of Schliemann's bank."[100]

The real reason is given in two letters written by Davidson to Rothschild and one written by May, all three dated the same day, 17 April 1852, presumably in order to catch an outbound steamer. The first is a composite letter in part of which Davidson indicated he had cancelled several drafts for Schliemann, totaling $12,075 (or $12,588 at 4 percent) and later wrote:

> We advise you above of a number of drafts upon London and N. York being cancelled: these Bills were forwarded to Mr. H. Schliemann in Sacramento for negotiation, but upon our discovery that he was acting fraudulently towards the public we immediately recommended his leaving the Country and closed our connexion with him.

He used some of the same words in the second letter, in which he wrote:

> For the last 7 months I have had an Agent in Sacramento for the purchase of Gold dust and Sale of Bills, Mr Henry Schliemann. I discovered however a short time back that he was systematically defrauding the public, and immediately proceeded myself to Sacramento and compelled him to leave the Country forthwith. I mention this circumstance to you because it is possible he may call upon you, altho' I do not suppose he will have the insolence to do so.

May, in the third letter, written in German, said that he had given a draft for £900 to Schliemann, who had left on 8 April. He had bought gold dust and sold drafts for Davidson and May in Sacramento since the previous September.[101]

> We finally heard, however, that he treated the public dishonestly and gave false weight, wherefore Davidson went immediately to Sacramento, took the business away from him, and induced him to leave at once, which he immediately did; he is believed to have made a good deal of money (*ziemlich vieles Geld*) through this swindling (*Schwindelei*) in Sacramento during this period.

Neither Davidson nor May spelled out the precise nature of Schliemann's fraud, aside from giving false weight,[102] but it was sufficiently serious for him to leave the country at once, presumably in order to avoid arrest and prosecution and perhaps less formal punishment from those he had defrauded. The episode of the lynching of Frederick Roe, which had occurred in February 1851, was probably well known to Schliemann.[103] Both Davidson and May stressed that he cheated the public rather than themselves. Indeed, he received a draft for £900 (dated 17 April and listed on the "Notes of Drafts" cited previously), and Davidson gave him a letter—the only known letter from Davidson in the Gennadeus archive, with an endorsement in Russian on the verso—dated 7 April 1852 and saying that he had examined Schliemann's account and found it correct: "I hereby acknowledge having received from you the balance of the account as well as all Bills of Exchange which I had deposited with you for Sale [presumably those he later cancelled], thus closing our account definitively to this day."[104]

Davidson, working with Schliemann's clerks, Grim and Satrustegui, set to work immediately to control the damage and cover up the reason for Schliemann's departure. The Sacramento *Daily Union* for 7 April, under "General Notices," included a statement, signed "Henry Schliemann & Co." and dated 7 April, that "The Banking business hitherto carried on in Sacramento City by Henry Schliemann & Co. is this day transferred to, and will henceforth be carried on by, B. Davidson, of San Francisco, to whom all the deposits have been delivered," with an addendum, signed "B. Davidson" and dated "Sacramento, April 7, 1852," saying that "With reference to the above advertisement, B. Davidson begs to notify that he will respond for all such deposits as have been transferred to him by Messrs. Schliemann & Co." The following day, after a repetition of the previous notices, a further notice, signed by Davidson and dated 8 April, said that "The Banking business hitherto carried on by Henry Schliemann & Co. will in future be conducted by Messrs. Grim and Satrustegui, who are authorized to act as agents for B. Davidson, of San Francisco, for the purchase of Gold Dusts. Drafts at par on San Francisco. No deposits taken." On the same day the *Daily Union* commented that "Davidson's Banking House in this city, we perceive is to be conducted in future by Messrs. Grim and Satrustegui, in place of Mr. Schliemann, who has been called to Europe to attend to some family matters of importance to him" and went on to wish Grim "the most abundant success." There was apparently no scandal and no break in the daily rhythm of the bank.[105] John Luck, in his letter to Rothschild of 26 April, wrote that "The great bulk of the gold remitted to you is purchased at Sacramento City by the agents of Messrs Davidson and May."

This stress on continuity, and perhaps also the fact that Davidson and May said that Schliemann cheated the public and that his accounts with Davidson were in perfect order, were significant because this episode came at a particularly difficult moment in Davidson's relations with the Rothschilds. The Paris house seems to have always treated Davidson with less confidence than the London house, and on 1 October 1851 Nathaniel, perhaps acting for Baron James, wrote from Paris to London that "It would be advisable to send somebody to San Francisco."[106] Some time later it was decided to send John Luck, about whom little is known except that he was a trusted agent of the Rothschilds, to check on their interests in America and Australia. His itinerary from St. Thomas (23 February 1852), San Francisco (16 April–24 May), and Sydney (21 August) to Melbourne (6 September) can be traced from his letters.[107] Davidson wrote on 4 April that he was expecting Luck and would assist him in all ways and on 17 April—the same day as the three letters concerning Schliemann were dated—that he had arrived.[108]

In the letter of 4 April Davidson thanked the Rothschilds—probably sarcastically, in the light of this and later letters—"for the very flattering proofs you give me of the opinion you have of my integrity and capacities. You write to us constantly to make you large shipments of Gold and to exert ourselves more in the prosecution of your business." He went on to warn them against mining companies, calling them "fraudulent schemes" and "swindlers," and advised against shipping quartz in view of the cost. He would do so if they wished "but I will be *in no way responsible* for the results and execute your order entirely at your cost and risk." This letter was written at almost the same time as Davidson heard the news about Schliemann and went to Sacramento to deal with it. Not long afterward he fell ill. A secretary wrote on 5 May that his health was improving, and he himself on 15 May referred to a "rather ... severe illness."

In the middle of these problems, on 14 April, he wrote his sharpest letter to date to the Rothschilds, replying to their letter of 8 April and "a letter of the same date from my Brother written at your instigation."[109] He replied in particular to two grievances. The first was that he had charged the London and Paris houses a 2 or 3 percent commission. To this "heinous offense," as he called it, now no doubt sarcastically, "I at once plead guilty," saying that he did so "in accord with your repeated instructions" and that they had always received the full amount he charged. "If the party referred to above [whose identity is unknown] or any one else asserts the contrary—*they lie*." He also pleaded guilty to the second grievance, that he had charged Mr Adnams $\frac{1}{2}$ percent for storing his gold dust,[110] and explained that since his object was "to induce all parties to dispose of their dust" he always charged to store it. "I have neither the time nor the inclination at present to discuss the merits of Mr Adnams' assertion that 'if a Banking business were well conducted it would yield immense profits.' Mr A would do far better to attend to the sales of his Saddlery and not to meddle with what he does not understand." He objected particularly to the Rothschilds saying "that I have robbed you of commissions," to their calling him a correspondent rather than an agent, and to their repeatedly upbraiding him "with the loss of your house by fire" and the affair of the iron houses.

> Do you then suppose that I am so devoid of all feeling that I am not possessed of any self respect, and that I will crouch beneath your insulting insinuations and accusations, and yield blindly to every exigency of yourselves and of the Clerks acting under you. . . . You lend a willing ear to every stranger who calls upon you, and you look on all my statements as so many falsehoods.

He did not mind the charge of imbecility, he said, "but when you accuse me of open robbery and of dividing the spoils with another, you have touched a more sensible chord, and I will yet one day prove to you how sadly you have been deceived with regard to myself."

Davidson returned to his defense (or attack!) in his letter of 28 June, written in reply to the Rothschild letter of 30 April, which was certainly written before they received his letter of 14 April. In it he again expressed astonishment "that you lend such a willing ear to everyone who returns from this country" and said that "in reply to your remarks [I] can only refer you to my previous communications." He would gladly reply to any questions, he wrote, "but I will not condescend to pay any attention to the idle reports which have been communicated to you by a certain young man," who is again not identified. In reply to the apparent charge that Adams & Co. and Page, Bacon shipped more gold than he did, Davidson said that they shipped for others as well as themselves and that they had agents all over the state, which was risky. Unless he was specifically instructed to do so, he said, he refused to engage in "a large transmissal business" and "every sort of swindle." He asked them in conclusion not to write constantly about his "want of activity and zeal" and not to criticize him so much.[111]

Between these two letters from Davidson, of which the effect is unknown, Rothschild received the letter of 26 April from John Luck (Appendix), who tried to pour oil on the troubled waters, since he dealt with several of the issues that disturbed them, including the ½ percent charged for storing gold, the quantity of gold shipped by Davidson and May, Davidson's real estate holdings, and the long-standing issue of the iron houses. On the whole his letter is very favorable to Davidson, though he admitted Davidson may have used bad judgment, or had bad luck, in some specific undertakings. He considered the expenses of the establishment "very modest indeed" and praised the caution of Davidson and May, who were "conducting your affairs both well and faithfully," and who disliked the "'suspicions' which they say are contained in almost every letter they receive from London." He himself, he said, had tried not to offend "or create the least suspicion as to the real object of my visit."[112] He then remarked on business conditions in San Francisco and California generally, of which he had a favorable opinion, in spite of some reservations, especially with regard to mining companies, on which he agreed with Davidson.[113]

This letter forms a suitable conclusion for Davidson's side of the story, as Schliemann's departure clearly ruffled no feathers in London, which was happy as long as the shipments of gold continued. Davidson stayed in San Francisco for another ten years as a prominent banker and respected citizen.[114] From 1857 until his departure in 1862 he was the consul of Sardinia (later Italy), owing probably to Rothschild influence in Turin but perhaps also to Davidson's French connections and the alliance between Napoleon III and Victor Emmanuel.[115] The official diploma, dated 20 March 1857 and signed by Victor Emmanuel and Cavour, and other documents among the Davidson papers show that he was consul and not, as has been said, acting consul.[116] He was also a founder of the French savings bank and protective association and was active in the city's French colony.[117] There are two letters in the library of the California Historical Society in San Francisco from Davidson to Senator Milton Latham, both beginning "'My dear Latham."[118] The first, dated 20 October 1861, is concerned with current affairs and conditions and concludes "with best love to your Wife." The second, dated 9 January 1862 (the year Davidson left San Francisco) is concerned with the war, economic affairs, amateur theatricals, real estate, and other matters. Davidson wrote that he had sold "his place" for $15,000 and that "The Jews are going to build a synagogue on the ground, which they look upon as already consecrated! by the purity of the life of the present occupant."[119]

After he returned to Europe, Davidson lived in Paris, where he was in 1864, and then in London, where he lived first in Piccadilly and later in Albert Terrace, before buying a house named Richmond Lodge, later renamed Sidholme, in Sidmouth, Devon. During 1865–67 he appeared occasionally in the correspondence of Baroness Lionel de Rothschild, who considered him an eligible bachelor.[120] He married some time after

1871 and had three children.[121] Meanwhile his bank in San Francisco continued under the successive names of Davidson and May (1862–63), Davidson and Berri (1863–66), and Davidson & Co. (1866–78), after which it was made over to Albert (or Alfred) Gansel (Ganse) and Jeffrey Cullen.[122] An undated clipping (probably 1862, when Davidson left San Francisco) from an unidentified newspaper found in the Davidson papers said that

> B. Davidson, senior partner of the long-established and staunch banking firm of B. Davidson & Co., has withdrawn and retired from the business. The bank is still conducted under its former style and title by the other partners who have been associated with it from the commencement. Its credit has always been of the highest rank.[123]

Davidson died on 21 September 1878 and was described in an obituary in the *San Francisco News Letter and California Advertiser* for 5 October 1878 as an "excellent gentleman" and a "genial, honorable, kindly gentleman and friend."

Schliemann's life after he left Sacramento is comparatively well known.[124] He went first to St. Petersburg, where he arrived in August 1852 and in October married Katerina Lyshen, whom he divorced in 1869, when he also obtained American citizenship. Later he moved to Greece and engaged in the archaeological excavations that brought him worldwide fame and honors, though recent research has established that he did not give up his propensity to falsehood and deception. According to Traill, "The question therefore is no longer *whether* but rather *to what extent* we should distrust Schliemann's archaeological reports."[125] He apparently revisited Sacramento in 1865, but it is not known whether he thought at that time of his first stay there, of his partnership with Davidson, and of his reasons for leaving. His secret was certainly well kept. In his obituary in the *Sacramento Bee* for 30 December 1890 his former agent Lloyd Tevis (by then president of Wells Fargo) described Schliemann as "a successful painstaking young fellow of about four-and-twenty years, who dealt fairly with all who sold or bought gold dust" and who "made a good sum of money" in Sacramento. The botanist H. W. Harkness, president of the California Academy of Sciences, although expressing some doubts about Schliemann's discoveries, described him as "a very clever man, who had the faculty for making everything he touched a success."[126] The fact that in his journal Schliemann falsified the record in order to cover his tracks suggests that he may have felt some shame. His powers of self-deception, however, were probably equal to his powers of falsification, and he may have looked back on this period when he laid the foundation of his fortune, and on his relations with Davidson, with relative equanimity and pride.

# Notes

1. Text cited here as RA (Rothschild Archive). Davidson's letters are found principally in XI/38/81 and XI/38/82. Other relevant documents are found in XI/109/81 and XI/121/7A, the General Letter Book (XI/148/112-13), and the American Letter Book. The T series, which will be referred to, includes copies and translations from other series, not all of which are now available. Relatively few letters sent from the Rothschilds to Davidson are known, but their contents can often be deduced from his replies. Davidson's letters are referred to here only by date. Insofar as is possible they are cited as they were written (including spelling, capital letters, word divisions, and underlining, printed as italics) aside from the expansion of abbreviations, many of which are inconsistent ("ab" and "abt" for *about*, "p" and "pr" for *per*, "pc" and "pct" for *percent*, etc. ) and would be confusing to reproduce. Most abbreviations are obvious or still in use, such as "comms" for *commission*, "dft" for *draft*, "m" for *thousand*, "prm" for *premium*, and "remce" for *remittance*. As a rule "Rothschild" refers to the company and "the Rothschilds" refers to members of the family.

2. Cited here, respectively, as CHS (California Historical Society) and Davidson Papers, which are the property of his great-grandsons, John Davidson Constable and the author of this volume, and of which there are copies in the RA.

3. Numbered BBB 11. These letters are also cited here by day and month only, without the year, because they all date from between September 1851 and April 1852. Like Davidson's letters, they are cited here as they were written. Schliemann's letters have been numbered and inventoried by Stefanie A. H. Kennell, who created the transcripts on which much of the Schliemann material in the present work is based.

4. Shirley H. Weber, ed., *Schliemann's First Visit to America 1850–1851* (Cambridge, MA, Gennadeion Monographs, 2; 1942), cited here as *Journal*. On this work see Hartmut Döhl, *Heinrich Schliemann Mythos und Ärgernis* (Munich-Lucerne, 1981), 8 ("Tagebuch mit Schilderung auch fiktiver Ereignisse") and, more generally, 76–78 ("Schliemann und die Wahrheit"); and David Traill, "Schliemann's Mendacity: Fire and Fever in California," in *Classical Journal* 74 (1979), 41–49, which shows that

Schliemann's accounts of the fire in San Francisco and of his own illnesses were fabrications, as were the reasons he gave for leaving Sacramento in 1852. On Schliemann's later life, in addition to his own works (cited in n. 19), see David Traill, *Excavating Schliemann* (Atlanta, 1993); Hervé Duchêne, *Golden Treasures of Troy: The Dream of Heinrich Schliemann*, trans. Jeremy Leggart (New York, 1995); David Traill, *Schliemann of Troy: Treasure and Deceit* (New York, 1997); Susan Heuck Allen, *Finding the Walls of Troy: Frank Calvert and Heinrich Schliemann at Hisarlik* (Berkeley, 1999); and on his reputation Duchêne, *Golden Treasures*, 130–35; and Suzanne Marchand, *Down from Olympus: Archaeology and Philhellenism in Germany, 1750–1970* (Princeton, NJ, 1996), 118–24, 148–50. See also the citations in n. 23.

5. Only one letter from Davidson to Schliemann (which will be discussed later) is known to survive.

6. The label inside reads "Manufactured by Cowan & Co., Edinburgh," with "quires" underneath.

7. On this technique, see Stefanie A. H. Kennell, "The Schliemann Letters: Artifact to Database," in *American School of Classical Studies in Athens Newsletter* (Summer 2002), 5. Cf. Stefanie A. H. Kennell, "Schliemann and His Papers. A Tale from the Gennadeion Archives," in *Hesperia. The Journal of the American School of Classical Studies at Athens* 76 (2007), 785–817, esp. 786–88.

8. The sixty-seven letters not written to Davidson were addressed to twenty other correspondents, including Adams & Co. (3), Baring Bros. (14), August Belmont (3), George Montgomery (9), Miguel de Satrustegui (Schliemann's clerk; 3), Henry Whetstone (Whitestone; one of Schliemann's agents, on whom more below; 7), Charles Young (4), and to Schliemann himself (by his clerk when he was ill). There is an index of recipients at the beginning of the volume.

9. The details about Meyer Davidson's early life are derived from his petition for British nationality in the RA XI/112/20, where he used the

spelling "Meyer." On his role when he was in Amsterdam financing the war against Napoleon, see Richard Davis, *The English Rothschilds* (London, 1983), 31–32; Niall Ferguson, *The World's Banker: The History of the House of Rothschild* (London, 1998), 94–95, 97, 105, where he is confused with his son Benjamin (111, 114, 116); and Herbert Kaplan, *Nathan Meyer Rothschild and the Creation of a Dynasty: The Critical Years 1806–1816* (Stanford, 2006), 86–89, 106, 116, and s.n. in index, who spelled his name both "Meyer" and "Mayer" and called him the "principal clerk" of N. M. Rothschild, whom Davidson had "the maturity and the courage to reprimand . . . for his destructive and deplorable behavior." Meyer Davidson died in 1846.

10. Jesse Cohen's name is also spelled Jessie, Jessi, Jesi, and Jessy. She lived from 1795 to 1869: see Kaplan, *Nathan Meyer Rothschild*, 6, 17, nn. 28, 30; and George Ireland, *Plutocrats: A Rothschild Inheritance* (London, 2007), 42–43. Jesse Cohen inherited £3200 from her father.

11. Among other works, see Victor Gray and Melanie Aspey, eds., *The Life and Times of N. M. Rothschild 1777–1836* (London, 1998), which has a chapter (by Melanie Aspey) on Hannah Cohen and her sisters. Benjamin Davidson wrote a letter of sympathy on the death of his aunt Hannah in 1850 (31/X/50), one of the few personal notes among his letters.

12. On James, see Anka Muhlstein, *Baron James: The Rise of the French Rothschilds* (New York, [1982]). Nathan Mayer was the only one of Mayer Amschel Rothschild's sons not to use the title "Baron" and the particle "de" or "von," which they (and their descendants) were granted by the emperor for their services to Austria during the Napoleonic wars.

13. See in particular Rainer Liedtke, "Kommunikationswege und Informationstransfer im europäischen Privatbankwesen des 19. Jahrhundert." PhD diss., University of Giessen, 2003, who calls the Davidsons a classic example of the Rothschild use of family members as agents (120). Davidson's older brother Lionel and younger brother Nathaniel both worked for the Rothschilds (see n. 16). According to Ireland, *Plutocrats*, 333, "Cousins who did not bear the Rothschild name had their place, but at the same time were expected to know their station."

14. On this mission see Ferguson, *World's Banker*, 470, and especially Liedtke, "Kommunikationswege," 189–209. According to his brother David's record book in the Davidson papers, he returned from Paris on 9 February and left from Folkestone on 19 February. His itinerary can be traced in his letters in the RA XI/38/81B (folder for 1847), including an interesting account of the near-loss of the gold and loss of a carriage, for which Davidson feared his employers would hold him responsible.

15. See his letters in RA XI/38/81B (folders for 1848 and 1849) and David Davidson's record book in the Davidson papers. Rothschild had interests in quicksilver (which was important for refining gold and silver) in both Spain and Chile: see Miguel A. López-Morell, *La casa Rothschild en España (1812–1941)* (Madrid, 2005), 137 (referring to Davidson, misspelled "Davison"); Stanley Weintraub, *Charlotte and Lionel: A Rothschild Marriage* (London, 2003), 15; the Rothschild Archive booklet on *Rothschild and Latin America* ([London], 2005), which cites one of Davidson's letters from Valparaiso; and Christian Platt, " 'Spanish Quicksilver': A Preliminary Note. The London Market, Global Trade and the Rothschild Monopoly (1830–1850)," in *The Rothschild Archive. Review of the Year April 2010 to March 2011*, 38–48. See also, generally, John Mayo, "Joshua Waddington and the Anglo Chilean Connection," in *Boletín de la Academia Chilena de la Historia* 71 (2005), 197–98. See also n. 41.

16. Lionel (born 1 October 1819) was in New York in 1839 and in Mexico from 1843 until 1852, when he was replaced by his brother Nathaniel (born 9 May 1833), who remained in Mexico until 1872: see Ferguson, *World's Banker*, 389, 394; Liedtke, "Kommunikationswege," 32, 210; Rothschild Archive booklet, *Rothschild and Latin America*; Alma Parra, "Mercury's Agent: Lionel Davidson and the Rothschilds in Mexico," in *The Rothschild Archive. Review of the Year April 2007 to March 2008*; Platt, "Spanish Quicksilver," 46, n. 2, who calls Davidson "Rothschild's plenipotentiary in Mexico for quicksilver"; and the forthcoming work of John Mayo.

17. RA XI/38/76B. I am indebted to John Mayo for bringing this letter to my attention. News of the discovery of gold in California first reached

Chile in May 1848, according to H. W. Brands, *The Age of Gold: The California Gold Rush and the New American Dream* (New York, 2003), 48.

18. Ferguson, *World's Banker*, 576, said that Davidson was "sent" to the West Coast from Mexico; and Weintraub, *Charlotte and Lionel*, 188, said that Nathaniel Davidson represented Rothschild in San Francisco.

19. Heinrich Schliemann, *Selbstbiographie*, ed. Sophie Schliemann, 4th ed. (Leipzig, 1942), 26; Heinrich Schliemann, *Troy and Its Remains*, ed. Philip Smith (London, 1875), 7 (autobiographical notice dated from Paris 1868); Heinrich Schliemann, *Ilios: The City and Country of the Trojans* (New York, 1881), 12; Traill, *Schliemann*, 22–23; and Duchêne, *Golden Treasures*, 25–26. See also the letter dated 28/XII in the letter-book of letters to Davidson, in which Schliemann referred to "the 5 years of my establishment in St. Petersburg."

20. Heinrich Schliemann, *Briefwechsel*, ed. Ernst Meyer, 2 vols. (Berlin, 1953–58), I, 45–47 (quote on 45).

21. Ibid., I, 47–48 (Schliemann to his father from St. Petersburg 20 July 1850). See the obituary notice in the *San Francisco Alta California*, 25 May 1850, calling him "Louis Schliemann . . . formerly of Germany, late of New York city, aged 25 years." Ludwig is mentioned in Schliemann's obituary in the *Sacramento Bee* for 30 December 1890 (see p. 59) as having been buried in the Yerba Buena Cemetery.

22. He was among the passengers listed in the *San Francisco Alta California*, 3 April 1851 (p. 2, col. 6) and was among those with letters waiting for them at the express office as indicated in the *Daily Transcript*, 5 June 1851, according to Charles Duncan, "Gold Rush Banker Achieved Renown for Troy Discovery," *Sacramento Bee*, 22 December 1974. The date of his arrival is significant in view of the fact that (probably in order to claim American nationality by having been in California when it became a state) Schliemann later said that he had arrived in California early in 1850.

23. On Schliemann in Sacramento, in addition to Traill, "Schliemann's Mendacity" and Duchêne, *Golden Treasures*, 26–28, see the more popular

articles, based for the most part on Schliemann's journal, by William L. Roper, "Sacramento's Mysterious Stranger," in *California Highway Patrolman* 29, no. 12 (Feb. 1966), 15, 51–53, 55–56; Duncan, "Gold Rush Banker"; and John F. Wilhelm, "Heinrich Schliemann's Sacramento Connection," in *California History* 63, no. 3 (Summer 1984), 224–29 (reprinted in *Sacramento County Historical Society: Golden Notes* 30, no. 4 [Winter 1984], 18–29), which has some interesting illustrations. Other illustrations of Sacramento in the 1850s can be found in J. Horace Culver, *The Sacramento City Directory* (Sacramento City, 1851; repr. Sacramento, 2000), which includes (14 ) a photograph of the sidewheeler *New World* docked at the foot of K Street in 1850 (Figure 15); Edmund L. Barber and George H. Baker, *Sacramento Illustrated* (Sacramento, 1855); part VIII ("An Instant City: Sacramento") of the *Gold Rush Exhibit* at the California State Library. On Sacramento generally in this period, see Theodore T. Johnson, *Sights in the Gold Region and Scenes by the Way.* 2nd ed. (New York, 1850), 129, 210, 218–19; William R. Ryan, *Personal Adventures in Upper and Lower California in 1848–9*, 2 vols. (London, 1850), II, 160–62, 218–19; Alonso Delano, *Life on the Plains and Among the Diggings: Being Scenes and Adventures of an Overland Journey to California*, originally published in 1854 and reprinted as *On the Trail of the California Gold Rush* (Lincoln, NE, 2005), 251–52 and 288–89; Mark A. Eifler, *Gold Rush Capitalists: Greed and Growth in Sacramento* (Albuquerque, 2002), including 58 and 185 on Front St. and the Waterfront, but with no reference to Schliemann; and Brands, *Age of Gold*, 273–74. The year before Schliemann arrived, Sacramento suffered from an epidemic of cholera: Charles E. Nagel, "Sacramento Cholera Epidemic of 1850," in *Sacramento County Historical Society: Golden Notes* 4, no. 1 (October 1957), 1–8.

24. *Journal*, 56–63 (quotes on 56 and 62), of which 61–3 (for no known reason) are in Spanish (with a translation on 98–100). This is followed by his fabricated account of the Great Fire in San Francisco (see n. 4). Much of this may have been written or revised subsequently in order to justify (at least to himself) his departure from Sacramento. For an almost contemporary description of Sacramento, see J. M. Letts, *California Illustrated including a Description of the Panama and Nicaragua Routes* (New York, 1853), 130–33, who said the city had 12,000–15,000 inhabitants and was very expensive.

25. Silver (and gold) could be sent either via Panama or round Cape Horn, which was cheaper and less risky but unpredictable. Letters could go via Panama or overland via New York. On the Cape Horn and Panama routes, see Brands, *Age of Gold*, 72–75, 93–121, and on the Panama route, Malcolm J. Rohrbough, *Days of Gold: The California Gold Rush and the American Nation* (Berkeley, 1997), 58–59.

26. August Belmont (Schönberg) (1816–90) appears frequently in both Davidson's and Schliemann's letters. He was the principal Rothschild agent on the east coast and an important financial and political figure in his own right: see *Dictionary of American Biography*, II (New York, 1929), 169–70; Liedtke, "Kommunikationswege," 147f; and the Rothschild Archive booklet *Rothschild in America* (London, 2005).

27. On San Francisco in the early 1850s, see Johnson, *Sights*, 210, 218–19; Walter Brodie, *Pitcairn's Island and the Islanders in 1850*, 2nd ed. (London, 1851), 219–22 (and 226–31 on the Panama crossing); and Delano, *California Gold Rush*, 356–71. On private coinages of "slugs," as they were called, see *The Works of Hubert Howe Bancroft*, XXIV: *History of California*, VII (1860–1890) (San Francisco, 1890), 165; Benjamin C. Wright, *Banking in California, 1849–1910* (San Francisco, 1910), 7–9, 161; Leroy Armstrong, "The Banking Record of Early Days," in *Financial California: An Historical Review of the Beginnings and Progress of Banking in the State*, eds. Leroy Armstrong and J. O. Denny (San Francisco, 1916), 77–81; Ira B. Cross, *Financing an Empire: History of Banking in California*, 4 vols. (Chicago, 1927), I, 127, 131, 133; Fred Marckhoff, "The Development of Currency and Banking in California," in *The Coin Collector's Journal* 15, no. 3 (1948), 60–61; and Otis E. Young, Jr., *Western Mining* (Norman, OK, 1970), 123–24. In December 1851 Davidson was the fifth from last of the 128 signers of the petition to Congress to establish a mint in San Francisco: CHS MS 897. A branch mint was opened in April 1854: Bancroft, *Works*, XXIV, 167; and Wright, *Banking*, 161.

28. Davidson returned to this issue in several letters written in 1850: "a great many Mexicans and other foreigners" had to pay $20 a week for the right to work and their departure may affect the supply of gold (31/

V/50); the Mexicans and Natives are badly treated, "altogether the boasted 'liberty' of the Americans has, so far as can be judged from this Country, no signification whatsoever" (15/VII/50); the Mexicans are "driven out of the Country" (30/IX/50); gold is less well washed and mixed with more sand owing to the departure of Mexicans (30/XI/50). On antiforeign agitation in San Francisco in 1849 and early 1850, see Rodman W. Paul, *California Mining: The Beginning of Mining in the Far West* (Cambridge, MA, 1947), 111, 202; Jay Monaghan, *Chile, Peru, and the California Gold Rush of 1849* (Berkeley, CA, 1973), 163–70; and Rohrbough, *Days of Gold*, 89, cf. 229 on the belief that access to the gold belonged to the Americans. On the prejudice against and decline of Native Americans, see Delano, *California Gold Rush*, 295–320; and Eifler, *Gold Rush Capitalists*, 28.

29. There is an unsigned and partially illegible copy of this letter, which may have been written by Anthony de Rothschild, in the American Letter Book, 94.

30. This was probably the lot on California Street, see p. 28 and n. 58.

31. See Josiah Royce, *California from the Conquest in 1846 to the Second Vigilante Committee in San Francisco* (1885; repr. New York, 1948), 298–301, on the appearance of the city in 1849 and on the fires in December 1849, 4 May 1850, 14 June 1850, and 17 September 1850. See also Brands, *Age of Gold*, 74, 254–59, 349–50, 363.

32. Frank Soulé, John Gibson, and James Nisbet, *The Annals of San Francisco* (1855; repr. Berkeley, CA, 1998), 512 (one of the five banks in San Francisco in 1849); Bancroft, *Works*, XXIV, 160; Theodore Hittell, *History of California*, 4 vols. (San Francisco, 1898), III, 443 (the third bank, after Naglee and Sinton and Burgoyne & Co.); Oscar T. Shuck, *Historical Abstract of San Francisco*, I (San Francisco, 1897), 84; Cross, *Financing*, 51 (dating the opening of the bank in February 1847, before Davidson reached San Francisco), 64, 67; Wright, *Banking*, 16; Armstrong and Denny, *Financial California*, 40; Marckhoff, "Development," 55 (one of the first seven banks), cf. 57; Dwight Clarke, *William Tecumseh Sherman: Gold Rush Banker* (San Francisco, 1969), 7 (one of the first seven

banks); Robert J. Chandler, "Integrity Amid Tumult: Wells Fargo and Co's Gold Rush Banking," in *California History* 70 (1991), 261.

33. Quoted from the San Francisco *Evening Bulletin*, 29 December 1855, in Cross, *Financing*, I, 64. On the banker and journalist James King, see Brands, *Age of Gold*, 390–91.

34. Charles P. Kimball, *The San Francisco City Directory . . . September 1, 1850* (San Francisco, 1850), 33. Copies of this and later directories were consulted in the archives of the CHS in San Francisco and the California State Library in Sacramento.

35. *The San Francisco Business Directory for the Year Commencing January 1, 1856* (San Francisco, 1856), 35, listed Davidson's bank as one of twenty-seven banks that received deposits, collected and remitted money, purchased gold dust and bars, received dust for coinage and assay, and discounted commercial paper.

36. On May, see Ferguson, *World's Banker*, 576; and Liedtke, "Kommunikationswege," 144–46. Davidson referred to May's coming from New York in a letter dated 12 May 1850. May's first name is given as Julius in the *San Francisco City Directory* for 1860 and 1861.

37. Schliemann complained on 7 February 1852 that May's letters gave him trouble.

38. The monthly renewals are recorded in the General Letter Book 82, 225, 235, 307, 339, 351. In his letter of 14/V/52 Davidson acknowledged receipt of the March letter of credit. Cf. Ferguson, *World's Banker*, 576.

39. There are three original bills of exchange (one "second of exchange" dated 13 Jan. 1853 [see Figure 10, p. 24], and two "third of exchange" dated 20 November 1862 signed, respectively, by Davidson and by Davidson and May) in the CHS, MS 5/10 and 181. See Marckhoff, "Development of Currency," 58, on the three sets of exchange drafts, of which the first went to the bank designated by the sender, the second was for the depositor, and the third was kept by the bank making the draft.

40. See the figures given by Soulé, Gibson, and Nisbet, *Annals*, 513, who said that gold circulated as currency at $16 an ounce but was often offered for less owing to the scarcity of coin and that at the end of 1849 banks paid between $15. 50 and $15. 75 an ounce for gold dust and $14.50–14.75 for quicksilver dust. Cf. Bancroft, *Works*, XXIV, 164, who indicates that the price fluctuated between $10 in 1849 and $17 in 1851; and Cross, *Financing*, I, 122.

41. Rothschild had a monopoly on Spanish quicksilver (see n. 15), the price of which fell from $114.50 a flask to $47.83 after the discovery in 1845 of the New Almaden mine in San Jose, south of San Francisco: see Johnson, *Sights*, 202; Paul, *California Mining*, 59, 63, 140, 272–77; and Young, *Western Mining*, 96, 102–20. For an interesting account of the New Almaden and Henriquita mines, see [James Mason Hutchings], *Scenes of Wonder and Curiosity in California* (San Francisco, 1861), 154–72. In 1857 Davidson gave evidence concerning the documents, which had been deposited in his bank, in the case of the United States vs. Andrés Castillero in the District Court for Northern California, of which there is a copy in the Special Collections of the University of California Irvine library and a microfilm in the Huntington Library. The case concerned the ownership of the New Almaden mine and sheds extensive light on the history of the mine. On the mercury process of refining, see Brands, *Age of Gold*, 237–39.

42. On quartz mining, which began in the summer of 1849, see Delano, *California Gold Rush*, 379–80; Bancroft, *Works*, XXIV, 173; Rohrbough, *Days of Gold*, 102, 200–02; and Banks, *Age of Gold*, 234–39.

43. Banks were forbidden to issue notes for circulation: Wright, *Banking*, 6–7. On the establishment of a mint in 1854, see n. 27.

44. On the establishment in 1850 of the assay office, see Bancroft, *Works*, XXIV, 167; and Cross, *Financing*, I, 134. There are other references to the assay office in Davidson's letters of 30/V/51, 31/VII/51, and 15/IX/51. The office did not have a monopoly on assaying, and Davidson and Schliemann continued to use Moffat & Co., which may have been cheaper though less reliable.

45. See n. 27 on private coinage, to which there are references in David-son's letters of 12/IX/49 (see Appendix), 4/III/51, 31/III/51, 14/IV/51, and 30/V/51.

46. See also Davidson's letters of 4/XII/51, 15/I/52, and 16/II/52 and several of Schliemann's letters, which will be discussed subsequently.

47. On these banks and their failure in 1855, see Cross, *Financing*, I, 68–71; Wright, *Banking*, 17, 21–22, 103–04; and Brands, *Age of Gold*, 348–50. In a letter of 14/IX/50 Davidson estimated that $3,500,000 worth of gold, at $16 an ounce, was exported monthly to the United States, Europe, South America, the English colonies, and China. The *Times* correspondent (24/I/51) put the figure for November 1850 at $4,337,000.

48. Chandler, "Integrity Amid Tumult," 261; see also Cross, *Financing*, I, 71–72.

49. "Continuation of the Annals of San Francisco: December 5, 1854, to June 3, 1855," in *Quarterly of the California Historical Society* 15 (1936), 272.

50. Clarke, *Sherman*, 110, 112, see also 337, where in 1858 Sherman wrote, "Davidson has the Jewish account." See n. 116.

51. Davidson's real estate ledger in the CHS, 88, 55, which runs from June 1853 until April 1862, lists expenses and rents for the Lafayette Building and the Clay Street lot only, which suggests either that he had disposed of the other real estate or that these were his personal properties, distinct from Rothschild real estate. The cash register lists among other income the Clay Street, Lafayette, Larkin, and Pearson rents. On rents in San Francisco, see Ryan, *Adventures*, II, 218–19.

52. Presuming that he is the same man referred to in a letter written from San Diego on 27 June 1853 by J. J. Ames and E. B. Pendleton to W. H. Davis, saying that "It is rumored her [sic] that you have disposed of your entire interest in *our* town to B. Davidson for $25,000," published

from the manuscript in the Pasadena Public Library by Andrew F. Rolle, "William Heath Davis and the Founding of American San Diego," in *Quarterly of the California Historical Society* 31 (1952): 41. There were other businessmen named Davidson in San Francisco at this time, but no one else with the initial B.

53. A. P. Nasatir, "A French Pessimist in California: The Correspondence of J. Lombard, Vice-Consul of France, 1850–1852 [cont. ]," in *Quarterly of the California Historical Society* 31 (1952), 313–14.

54. This is one of the relatively few references in Davidson's letters to his salary. His brother Lionel in Mexico, according to the American Letter Book, 119, was paid £1000 a year for salary and expenses.

55. See pp. 21, 30, and 33.

56. He again gave up his salary and said he would take whatever he was offered plus expenses. In his letter of 12/VII/50 he referred to the first expense money he had taken since leaving England.

57. Several of these matters are discussed in the letter of John Luck written 29/IV/52 (see Appendix).

58. The California Street property was not mentioned because it had been sold, Davidson reported on 31 July 1851, "as you appeared displeased at the idea of holding so much land in this town." George Luck's reference in his letter of 29/IV/52 (Appendix) to this sale and "the freight charges" suggests that Davidson's house from Valparaiso may have been erected here.

59. RA XI/109/73/2. See Ferguson, *World's Banker*, 512 and, on James's rudeness and obsession with money, see Muhlstein, *Baron James*, 159, 211. The gold trade was undoubtedly profitable, but exactly to whom and how Rothschild sold the gold is uncertain.

60. See Davidson's letters of 14/VI/51 and 14/XI/51, saying that business was dull. On 22/I/51 he asked them not to send bolts and sheet iron.

61. RA XI/38/76B. See also American Letter Book, 121, on the iron houses, with an account of Lionel's expenses, and John Luck's letter of 29/IV/52 specifying that they were "of Messrs Drusina & Co." According to advertisements in the London *Times* for 2/I/50 and 25/IX/51, they were priced at £12 and £20—Rothschild doubtless paid less— and the shipment may therefore have numbered upward of 1,000 houses.

62. There is an interesting description of the houses' construction, under the heading "Cast-iron Houses," in the *Trumpet and Universalist Magazine* for 6 April 1850 and in the advertisements in the *Times* for 2 January 1850, where they were described as 12-foot square, with iron swing bedsteads, and as fireproof, and 21 May 1850. I am indebted to Professor Carol Gluck of Columbia University for these and other references on iron houses.

63. See the London *Times* for 5 March 1850. The number of iron buildings in San Francisco on 1 September 1850 was estimated at 68 as reported in the San Francisco *Sun* and in the *New York Observer and Chronicle*, 27 October 1853.

64. He went on to say that "In summer they become as hot as ovens, and in winter as cold as wells. As to their being fire-proof, after the tests our conflagrations have put them to, the idea is laughed at."

65. On 3 July the London *Times* reprinted an article dated 15 May from the *New York Herald* saying that "Iron-houses curled up like sheets of paper before a fire." See Brands, *Age of Gold*, 256–57.

66. See also letters of 14/IV/51, 15/VII/51, 30/VII/51, 14/VIII/51, 14/XI/51, 31/XII/51, 16/II/52, 14/III/52, 17/III/52, and 30/X/52, where he said that he hoped to get $18,000–$20,000 for the remaining houses.

67. The population of California grew from 14,000 in 1840 to about 100,000 by the end of 1849, and 223,000 by late 1852: see Paul, *California Mining*, 20–25; cf. Rohrbough, *Days of Gold*, 19 (the population increased eight times in a year), 33. The arrival of emigrants sometimes created

problems for Davidson, like those who came from Paris whom Davidson was "utterly at a loss to recommend them what to do" (31/X/50). There is a letter of introduction dated 15/VIII/50 in the General Letter Book, 426, for a Mr. Samson "who is going out to California to try his fortune" and who later bought a note for £25 (31/XII/52, note no. 3584).

68. On the Vigilante Committees in San Francisco see Delano, *California Gold Rush*, 389–91; Royce, *California*, 328–33, 344–66; Paul, *California Mining*, 207; Philip J. Ethington, *The Public City: The Political Construction of Urban Life in San Francisco, 1850–1900* (Cambridge, MA, 1994), 86–169; Rohrbough, *Days of Gold*, 157, 159, 160; and Brands, *Age of Gold*, 265–68, 351–52, 376–78.

69. To these should probably be added two drafts for £602 and £140 dated 15 and 18 November, respectively, for "Schulte and Schliemann," though Schliemann is not known to have had a partner named Schulte.

70. This letter is copied out of place in the letter book and the date is uncertain.

71. *Journal*, 65–66. So far as is known, Schliemann never shipped gold directly to London, but it may have flattered his pride to suggest that he did.

72. There are copies of this advertisement in Traill, "Schliemann's Mendacity," 45 and (slightly different) in Wilhelm, *Golden Notes*, 21. Later advertisements are cited from photocopies of the *Daily Union* in the California State Library. I have found no reference to Schliemann's bank in any history of banking in California or Sacramento, including Eifler, *Gold Rush*, or in the three articles on early Sacramento banking in *Sacramento History: Journal of the Sacramento County Historical Society* 2, no. 3 (Summer 2002), aside from one reference in the caption to the illustration on 17.

73. See Schliemann's references to advertisements, signs, placards, circulars, and cards in his letters of 21 October, 6 November ("more than ten

plackards in and outside our office" and "the cards you got printed for us and which at your wish we give to every miner who comes in"), 7 November, 8 November, 29 January, and 12 February.

74. The $20,000 bond was kept in a special account (8/XI). Parts of this letter are blurred and difficult to read.

75. This return commission or rebate (usually ½%) was to prove a source of controversy. On 7 November Schliemann wrote, "Regarding our arrangement with you to share the return-commission of Moffat on gold deposited for our account in that establishment we find your *full* confirmation in your contract-letter. We never attempted to extend this new stipulation upon *past* and solely on future transactions, . . . By virtue of this document we can never again deposit anything in or have any intercourse whatever with the Mint." This shows that Schliemann was permitted to send gold or quicksilver dust on his own (as other evidence shows) or foreign (that is, for other people) account for assay by Moffat, provided he did so with Davidson's knowledge. He often in fact did so through Davidson.

76. For most of the period Schliemann's two clerks were Grim and Satrustegui, on whom, see pp. 5, 54–55, n. 105. According to the *Journal*, 67 (under 1 September 1851) Lewis Saynisch was added, but he does not appear in Schliemann's letters.

77. The "& Co." was apparently an informal, nonbinding recognition of the partnership. On 2 November Schliemann wrote, "Please let me know the reason why you wanted me to sign Hy S & Co. and not that old name under which I was born and baptized. I am under the impression that though I may sign Hy S & Co. yet in case of my death no person is legally empowered to take charge of the settlement of my business, unless indeed he can produce a written certificate of a company or that of partnership having existed betwixt him and your humble servant."

78. On 5 February Schliemann wrote that Davidson could "refuse our Gold any time it pleases you but we would then thank you for selling

such refused parcels at the best price you can get rather than to be shaved by M[offat] & Co."

79. See Schliemann's letters to Whetstone of 2 November and 10 November, where he warned him against the drafts of some San Francisco banks, expressed the hope "soon to receive a large amount of fine dust from you," and said that "The House of Rothschild does not ship any gold dust on foreign account, neither do we do so here," which suggests that Whetstone had proposed shipping gold himself. There are also several letters addressed to Whetstone in Schliemann's letter book (see n. 8).

80. On Lloyd Tevis (1824–99), see the *Dictionary of American Biography*, XVIII (New York, 1936), 384–85; and Robert D. Livingston, "Lloyd Tevis: Corporate Raider," in *Sacramento History: Journal of the Sacramento County Historical Society* 2, no. 3 (2002), 33–47, with a brief reference to Schliemann on 35. Tevis was a shrewd but unscrupulous man, perhaps not unlike Schliemann himself.

81. Because Davidson had sent only $2000 that day, Schliemann charged him $15 (1½% of the other $1000).

82. On 19 November Schliemann agreed to obtain himself the $3000 in small coin at a cost of $45 (1½%) to Davidson, but on 20 December he asked for $6000 in American coin "even if you should have to pay as much as 2½%."

83. Gregory's Express advertised in *Bogardus' San Francisco, Sacramento City and Marysville Business Directory, for July 1850*, 5, that it "sends a Messenger to the United States by every Steamer, and to Sacramento City daily." See Cross, *Financing*, I, 73–74, on Gregory's Express (which was taken over by Wells Fargo in 1852) and 67–74 on other express companies.

84. Cross, *Financing*, I, 68–71.

85. Schliemann refused, however, to pay Davidson a ½ percent commission "for having once reciprocated Adams's kind service on your behalf (by supplying us with small Gold)."

86. See also Schliemann's letters of 7/XI, 11/XI, and 29/XI.

87. In these letters Page, Bacon & Co. appears as an ally of Schliemann's, but shortly before, on 31 October, he complained that Page, Bacon had attracted many miners to its bank by displaying "large printed plackards as to the sailing of and passage money on the Panama and Nicaragua steamers" and asked Davidson to send him "some of these prints." On Page, Bacon & Co. in Sacramento, see Armstrong and Denny, *Financial California*, 136; Robert D. Livingston, "Banking Panic of 1855," in *Sacramento History: Journal of the Sacramento County Historical Society* 2, no. 3 (2002), 15–28; and James Henley, "Adams & Co. and Page & Co. in the 6th District Court," in *Sacramento History: Journal of the Sacramento County Historical Society* 2, no. 3 (2002), 29–32; and on Sacramento banks generally, see Cross, *Financing*, I, 75–86; Marckhoff, "Development of Currency," 64–65; and Robert D. Livingston, "Sacramento's First Bankers," in *Sacramento County Historical Society: Golden Notes* 30, no. 4 (1984), 1–6 (3–4 on Schliemann).

88. In spite of Schliemann's confidence, the price of gold all over town ten days later was $17^{1}/_{4}$ (5/XII/51). In fact (though Schliemann could not know it) the boom days of 1848–51 were over and gold production declined (and prices rose) after 1851: see Paul, *California Mining*, 116–17, 345–46. There is a vial of gold dust among the Davidson papers labeled "#6/Fort Yale/4.3.1858." Fort Yale is on the Fraser River in British Columbia, just north of the U. S.–Canada border. The gold rush there took place in 1858, and this gold was presumably sent to Davidson at the time. It suggests he was by then looking further afield for gold.

89. See Schliemann's letter of 6/IV, his last to Davidson.

90. On the price of land in Sacramento, see Delano, *California Gold Rush*, 251, who, writing in 1854, said the price rose from $200 to $30,000 in a year, and Eifler, *Gold Rush Capitalists*, 51, 54, 64, 121, 249, n. 55, who estimated the rise from $250 to $5000 a lot.

91. On the floods in Sacramento, see Johnson, *Sights*, 218–19; Rohrbough, *Days of Gold*, 162–63; and Eifler, *Gold Rush Capitalists*, 94–101, 129.

92. This was probably Grim. The first letter in his hand is dated 11 November. A salary of $250 a month ($3000 a year) is less than Davidson paid ($4000) his clerk in San Francisco.

93. See *Journal*, 69–71.

94. In his dispute with Davidson over shipping, Schliemann wrote that he occasionally shipped by Adams & Co "on our own account through them to you" (19/III).

95. Two days later, on 28 December, Schliemann wrote to say that "By mistake we put on the box with 1000 ounces Golddust value $4000—,whereas the costing price is only $17.250" and asked Davidson to put only 1000 ounces Gold, without a value, on the bill of lading.

96. See Schliemann's letters of 29/XI, 18/XII, 12/II, and 19/III.

97. Traill, "Schliemann's Mendacity," 46–47; Allen, *Finding the Walls*, 112.

98. Two weeks later, however, on March 6, a month before his departure, Schliemann wrote, "The northwind blows its bleak blasts through our muddy and lifeless streets and everything looks gloomy and dead." He also expressed his dislike of "foreigners" (Chinese, Mexicans, South Americans, Hawaiians, and especially American Indians) in the *Journal*, 68–89.

99. *Journal*, 74.

100. Traill, "Schliemann's Mendacity," 46–47.

101. The word for "drafts" is not easily legible, but the meaning is clear. I am indebted to Klaus Weber and Melanie Aspey for assistance in transcribing this letter.

102. On various forms of fraud, see Bancroft, *Works*, XXIV, 164, n. 26 (mixing with black sand, low-grade gold, false weight, spurious dust);

and Young, *Western Mining*, 45. Schliemann's fraud may also have been associated with his dealings with his agents, especially Lloyd Tevis (see n. 80), which he was at pains to keep secret from Davidson (see p. 40). Cf. Duchêne, *Golden Treasures*, 28.

103. Eifler, *Gold Rush Capitalists*, 203–12, and 188 on lynching generally.

104. In view of its value to Schliemann, some question could be raised about the authenticity of this letter, but it appears to be in Davidson's hand and was also of use to him for the good name of his bank.

105. Notices similar to the original one (all dated 7 April) appeared in several Sacramento papers throughout April. May, Grim, and Satrustegui advertised in the *Daily Union* under their own names as "Bankers, No. 3 J. street, near Front St. Agents for B. Davidson, San Francisco. Exchange for sale on New York, London, Paris, Frankfort, Hamburg, etc., etc." Marckhoff, "Development of Currency," 64, includes "A. K. Grim and F. Rumler" among the banks in Sacramento in 1853. Grim and Rumler, Bankers, also appear among the institutions in the Tehama Block (see Figures 4–6, pp. 11–13). This may have been the successor to Schliemann's bank but was probably a different company.

106. RA XI/109/79/4. See Ferguson, *World's Banker*, 576, no. 110, who cites this letter as from James; and Liedtke, "Kommunikationswege," 144–46 on the differing attitudes toward Davidson in Paris and London (which may go back to Mayer's instructions not to send gold to Paris), and 279–80 on the Rothschilds' distrust of their agents.

107. RA XI/121/74, which does not include the letter of 26 April published here in the Appendix.

108. Luck had a credit of £200: see American Letter Book, 95. On 30 July Davidson listed charges on J. Luck for four packages, including "Advertising in San Francisco 'Herald'," without specifying for what.

109. There is a problem with the dates here, though they are both clearly written, since Davidson could not have replied on 14 April to letters written (presumably in London) on 8 April.

110. Adnams is an uncommon name, and Luck apparently called him Adams in his letter of 26 April, but the spelling in Davidson's letter is clear.

111. On 15 July, however, Davidson wrote to say that the Rothschilds were wrong "in supposing that the business at every other house is conducted in a more advantageous fashion than your own," showing that the criticisms continued.

112. In this he was not successful. Davidson's comment in his letter of 17 April that Luck would "write to you whatever he pleases respecting my mode of conducting the Business here" shows that he was under no illusions concerning the purpose of Luck's visit. He also said that he had shown his letter to Luck.

113. Cf. the report (of which there is a copy in the RA XI/109/81) of Klaucke, Mel & Co., dated 15 May 1852, which also presents a rosy picture of the economic future of San Francisco.

114. "Davidson (Benjamin) & Co. (Julius May) bankers (and Consul for Sardinia)" is listed in *The San Francisco Directory* for 1861. In later volumes he is cited as "res London" (1862) and "resides London" (1865), when the company was listed as "Davidson, B. & Berri (Emanuel)." Mt. Davidson (on which Virginia City is built) was probably not, however, as believed in the family, named for him: Helen Carlson, *Nevada Place Names: A Geographical Dictionary* (Reno, 1974), said that it was named for Donald Davidson or George Davidson (92) or for a prospector who came to Carson Valley in 1851 (239).

115. On this appointment and the controversy surrounding it, see Ernest S. Falbo, "State of California in 1856: Federico Biesta's Report to the Sardinian Ministry of Foreign Affairs," in *Quarterly of the California Historical Society* 42 (1963), 312–13.

116. Falbo, "State of California," 332, nn. 6–8 (but not in the text); Alessandro Baccari and Andrew Canepa, "The Italians of San Francisco in 1865: G. B. Cerruti's Report to the Ministry of Foreign Affairs," in

*California History* 60 (1981/1982), 356. Contemporary sources, including *The San Francisco Directory* for 1858, 1860, and 1861, are unanimous in describing him as consul. In 1864, after Davidson's retirement, he was appointed a knight of the Ordine dei Santi Maurizio e Lazzaro in recognition of his services as consul in San Francisco (Davidson papers).

117. Baccari and Canepa, "Italians," 356. On the French in San Francisco, see in particular Abraham P. Nasatir, *French Activities in California: An Archival Calendar-Guide* (Stanford, 1945), where there is no reference to Davidson.

118. CHS MS 1245, 5 and 6. On Latham, who served as collector of the port of San Francisco before becoming senator, see Soulé, Gibson, and Nisbet, *Annals*, pt. II: Dorothy Huggins, *Continuation of the Annals of San Francisco, 1: From June 1, 1854, to December 31, 1855* (San Francisco, 1939; repr. 1955), 6, 12, 39, 51, 61, 64, 65. There is a letter book of Latham's from his time as collector of the port of San Francisco in the Huntington Library (HM 18953).

119. Aside from the references in the letters of William Tecumseh Sherman cited previously (nn. 32, 50), this is the only reference I have found to Davidson's connection with the Jewish community in San Francisco, but there may be others in later letters. The synagogue in question may be the Congregation Ohabai Shalome ("Bush Street Temple") established in 1863: see Martin Meyer, *Western Jewry: An Account of the Achievement of the Jews and Judaism in California* (San Francisco, 1916), 50–53.

120. These details, from documents in the RA, were kindly provided by Melanie Aspey.

121. Davidson's wife was named Olga. Her identity is unknown. She has attracted some attention because by her second marriage (to Adolph Lindemann) she was the mother of Frederick A. Lindemann (Lord Cherwell), Churchill's scientific advisor: see especially the Earl of Birkenhead, *The Prof in Two Worlds: The Official Life of Professor F. A. Lindemann, Viscount Cherwell* (London, 1961), 16, and Adrian Forb, *Prof: The Life*

*of Frederick Lindemann* (London, 2003), 2, 6–7. These works repeat the family story, doubtless derived from Olga, that her father was a Scottish or American engineer named Gilbert Noble who married in Russia (and met Davidson there in 1847–48) and went to America, where his daughter was born either in New York in 1850 or New London in 1851. After her father's death, according to this account, Davidson became her guardian, arranged for her education in Paris, and later married her. There is no evidence for this story, and no trace of the family in America. The most satisfactory clue to her identity, discovered by her grandson Noel Vickers, is in the British census of 1871, which lists, among the inhabitants of Davidson's house at 6 Albert Terrace, Olga O'Brien, as a "Visitor," "Married," born in Manchester, and aged 23, which would make 1848 her year of birth. Davidson himself is listed as "Unmarried" and as "Retired Bank Agent." There is no certainty that he married Olga O'Brien, but the name Olga is unusual. Davidson's wife was a gifted musician and composed at least four pieces for piano, one dedicated to Benjamin Davidson and another to Princess Maria Galatzine.

122. See Cross, *Financing*, I, 51. May left in 1863 and was replaced by Emanuel Berri, on whom see Falbo, "State of California," 332, n. 8.

123. Details of his will were given in the *Daily Alta California* (1, col. 3) for 2 June 1880 and in a clipping dated 1 June 1880 from an unidentified newspaper found in the Davidson papers; see also Shuck, *Abstract*, 84.

124. See in particular the works cited in n. 4.

125. Traill, *Schliemann of Troy*, 5.

126. There are two letters in the Schliemann archive in Athens dated 1881 from Tevis to Schliemann, who visited Tevis when he was in California. On Harkness, who lived from 1821 until 1901, see http://en.wikipedia.org/wiki/H. W. Harkness. On Tevis, see the references in n. 80.

# Appendix

I am indebted to Melanie Aspey of the Rothschild Archives for the transcripts of these two letters. The original of Davidson's letter was addressed to "Messrs de Rothschild Frères, Paris" and dated "12th September, 1849." A copy (marked "Duplicate per Steamer," addressed to "Messrs N. M. Rothschild & Sons, London," and dated "San Francisco Sept. 12, 1849") is in RA XI/38/81B. The differences between the two versions (except for abbreviations, capitalizations, paragraphing, punctuation, spelling, and underlining [of which there is none in the copy]) are indicated in the notes. It is hard to assess the significance of the differences. Many are minor revisions, clarifications, and omissions, some of them probably owing to carelessness. The omission at n. 65, for instance, is almost certainly owing to the common scribal error of jumping from like to like (in this case "house" to "house") omitting the words in between. The changes at n. 71 (omitting London as a place to insure the gold) and the omission at n. 76 may mark intentional changes, and the long omission at n. 73 is possibly significant, though why Davidson sent this passage to the Rothschilds in France but not in London is uncertain. He may have felt its contents were covered by "it would be very advantageous for several reasons." The double postscript was sent only to London and was presumably an afterthought. The original of John Luck's letter is not among his other letters in RA XI/121/7A, Box 52, but there is a somewhat inaccurate copy, with several lacunae, in T6/282. Much of the material in it is referred to in the text of the present work. On the Hartog affair, see Davidson's letters of 14 June, 14 August, 14 November 1851 and 15 January 1852.

From: Benjamin Davidson, San Francisco
To: Messrs de Rothschild Frères, Paris
On: 12 September 1849

Private

Gentlemen,

I have to apologize to you for the extreme brevity of my
letters by last mail, but upon my first arrival here I had so
many things to attend to, and was put to so much
inconvenience, in consequence of my not having any room
of my own, that I found it impossible to furnish a separate
and distinct account of the state of affairs here both to your
London house and to yourselves. I will however now
previous to the arrival of the steamer,[1] endeavour to explain
to you the position of affairs in this country, and draw your
attention to such transactions as I consider most likely to
meet your views.

The combined circumstances of inexhaustible gold mines
having been discovered and of this country having fallen into
the hands of an enterprising people like the Americans, have
produced a state of affairs here, which I suppose are
unparalleled in the history of the world, and which therefore
it is necessary to consider, without reference to and without
forming any comparison with what passes elsewhere.

The class of people that have been[2] suddenly attracted to this
country are of a most mixed nature, as may naturally be
imagined. That a considerable number[3] of them are men who
in other countries could[4] never succeed whilst having to
compete with the more honest and industrious classes
cannot be denied, yet at the same time no new country ever
attracted so many men of talent and enterprise as have been

and are still constantly arriving from all parts of the world, but more especially from the U. States.

The majority of the newcomers, being possessed of but very limited means, proceed as fast as possible to the mines in the interior.[5] Men who have before never handled a spade, or[6] been in any way habituated to manual labour, picture to themselves that they can resist any hardship in search of the golden harvest which some lucky few have reaped before them, but the greater number[7] of them greatly overrate their own powers of endurance, and may consider themselves fortunate, if after having collected a few ounces of gold, they still have sufficient strength left to bring themselves back to the point from which they started. At first the Americans did not interfere with the Indians and Spaniards, but they soon became jealous of their superior tact and success in working the gold, and have of late, greatly to the detriment of commerce, driven away and murdered numbers of Indians, Mexicans and South Americans.

This circumstance is to be regretted, in the first place because the amount of gold extracted will be considerably diminished, and secondly because the foreigners (as they are termed here) were the greatest consumers of goods, and in general very profuse in the expenditure of their easily acquired wealth.

As I have on a former occasion stated, it is utterly impossible to form any just calculation of the quantity of gold produced. I presume however that there is *for the present* not the slightest fear of any amount being exported sufficiently great to influence its relative value in Europe.

That a vast tract of country is more or less besprinkled with gold, is an undoubted fact, and I expect that very shortly it will be[8] found that the most advantageous way of working the gold will be by means of machinery, similar to that used in South America in the extraction of silver by means of quicksilver.

Indeed similar means are already being resorted to and are found to answer most admirably, and it is in consequence of this that the demand for quicksilver has arisen here.

I will make the quicksilver mines the subject of a separate letter, but will now call your attention to what I consider will be a most profitable[9] business after the coming winter.

I refer namely to the construction of the necessary machinery for working up that part of the sand which has once passed through the hands of the regular gold washers, but which still contains a large per centage of gold, which they in their rude mode of working could not extract. The present rate of labour is so high and the expense of transporting to the interior and constructing the necessary apparatus would be so great, that no man who is not possessed of ample means could undertake a similar enteprise.

Besides the regular gold washing, ([ . . . ][10] as they are called in Russia), a vein of gold ore has just been discovered by Colonel Fremont, and I have no doubt but that this discovery will soon give rise to others of a like nature in different parts of the country.

This gold ore must be crushed by machinery and amalgamated with quicksilver, and ere long we shall undoubtedly have companies formed here for that purpose. I am aware that capitalists in Europe are generally averse to having any connection with companies formed for undertakings of this nature and at so great a distance, but I consider that in this country a company well managed by parties *on the spot* would yield an immense profit.

Besides San Francisco, several other towns are being
established at different points of the bay, and it is considered
by some persons as being doubtful whether this place will
remain the principal port of California. However I do not
think there is any fear whatever of the site of the town being
changed, there is too much capital invested here already[11] in
buildings, and too many merchants established here[12] to
allow of any alteration being made at least for several years.
The Civil Governor, General Riley, resides at Monterey, and
General Smith, the Military Governor, is personally
interested in some towns in the interior. His influence will
however never[13] be sufficient to make any change[14] on this
head, and the ships must at all times come to that spot
where they[15] find the readiest market for their goods, and[16]
without any reference as to whether this is actually[17] the
most advantageous spot to disembark the[18] goods for the
consumption of the interior.

I have as yet not been into the interior myself, and from
what I hear of the different towns I do not suppose there is
any business likely to suit us at present. Gold dust is not to
be procured in the interior in exchange for gold coin at
lower rates than is the case here, therefore no advantage
could be gained by sending coin for its purchase.

The population of San Francisco may be estimated at about[19]
15,000 men, a large part of whom are at present dwelling in
tents, but houses are springing up with amazing rapidity in
every direction,[20] and every inch of land is already claimed
and held at the most exorbitant rates. The largest fortunes
that have been gained here have been accumulated by the
speculators[21] in building lots and houses which sell and rent
at perfectly fabulous rates.

Later it is probable that disputes may arise as to the validity
of the titles of the parties who have sold the land, but this

question cannot be decided until a regular government is appointed, and even then this[22] will not be an easy matter to regulate.

For the next 6 months the construction of houses will be a most profitable business, as one year's rent at the present rates would suffice to pay the whole cost of the construction and of the land.

Nearly the only article which continues to maintain[23] its price here is American lumber which sells at 325 and[24] 350 $ the m/feet,[25] leaving an immense profit to the shippers.

There are an immense number of merchants, money changers and shopkeepers already established here, but taking them as a body, they are the most unmercantile set it is possible to imagine, the majority of them are possessed of no means and sacrifice the goods which are consigned to them in a most shameful manner. The expenses are[26] enormously high and the market is[27] so overstocked that the better class of the merchants find the commission of 10%— high as it is—scarcely repays them, and the consequence is that they are no longer willing to receive[28] any consignments.

As yet none of the French vessels reported to have sailed from Bordeaux and Le[29] Havre have arrived, but the prospects for[30] the sale of wines and goods in general[31] are daily becoming more gloomy, and I fear, from the present state of market and of the immense expense of landing, storing etc,[32] together with the high duties imposed by the Americans on[33] most French produce, that the shippers will experience in many instances severe losses. Later I have no doubt but that certain current articles, which in the course of the next 3 months will be sold at any price, will again rise in value, although an improvement can scarcely be looked forward to[34] until the continued influx of ships should cease.

There are no French houses of any importance here, and it will be an easy matter[35] for me to keep the run of the market. I expect that a good business may be done[36] by buying up French goods and "liquides" which do not deteriorate by being stored. In American and English[37] goods there is not so[38] much chance of operating to advantage, on account of the trade being already in the hands of too[39] many.

I wrote to you last month that I thought shipments from France[40] of sundry articles might answer, it would however perhaps be more prudent to abstain from entering into any operations which would take so long a time to realize, particularly as this place in itself offers so many favorable opportunities for investments.

In consequence of the large capital that is being invested in houses, and the amounts that are[41] absorbed by the duties, money has become very scarce here and the rate of interest varies from 4 to 6 per cent per month, the security being I think quite as good as what can be[42] obtained in other countries.

A very good business may be done here in discounting, and likewise in the purchase and sale[43] of gold dust. At present the gold is worth from about[44] 15$\frac{1}{4}$ to 15$\frac{1}{2}$ $ per[45] oz. in exchange for coin, whereas the ounce is generally received in[46] payment at the rate of 16$.[47] This difference between the price[48] of gold dust and coin will naturally cease as soon as[49] a mint is[50] established here, and on this subject I should be particularly obliged to you if you would write to Mr Belmont or to your agents in Washington or Philadelphia, requesting them to inform me *immediately* it becomes known that the

Government is going to establish a mint here and about
what time the necessary machinery may be expected out
here, for as soon as the gold is coined on the spot, it is to be
supposed that its nominal value will rise at least 10%, and it
would therefore be a very advantageous and safe[51] business
to purchase a large quantity of the dust previous to the mint
being put into operation.

I beg to remark in general that it will be at all times most
essential[52] to receive regular advices from the U. States
concerning the intentions of the Government relative to this
country in order to make any favourable bullion operations.

At present a house here is endeavouring to force into
circulation a gold coin of their own manufacture,[53] and
should they succeed in so doing to any extent then[54] they
will gain very largely on[55] the transaction.

I do not know[56] the plan I am now going to propose to you
would be practicable, but should such be the case, and you
yourselves will be the best able to decide[57] upon that point,
we might likewise enter into an operation which would yield
a very great profit.

I propose namely[58] to establish a sort of bank, and to issue
notes for various amounts payable at sight in San Francisco
in some of the principal cities of the States and likewise at a
fixed exchange[59] in London and Paris.

These notes would be delivered here on receipt of the gold
dust, and I am convinced that coming from your house they
would be taken willingly not only in San Francisco, but
likewise all over the mining districts, as they would be much
more convenient to carry about and to remit than gold dust.

The profits arising from a bank here would be very great, in
the first instance[60] you would have the benefit of the

difference between the value of one[61] ounce of gold dust and
16$ currency,[62] secondly the use of the money for the
purpose of discounting, and lastly a decided gain from the
number of notes that would be lost, burnt or otherwise
destroyed and which consequently could never be
redeemed.

This will appear to you perhaps as being a very wild scheme,
but I am convinced that 6 months ago a bank with power to
issues notes would have yielded immense profits, and I still
think that, after a mint should be established, a plan
somewhat similar to the one I have mentioned might be
carried out with as great advantage as any undertaking in the
country.

I have already mentioned that the reason of the existing
difference between the relative value of [63] gold dust and coin
is[64] owing to the custom house only accepting the latter in
payment of the duties.

There exists at present a very large amount of money in the
hands of the directors of the custom house,[65] which is
another subject to which I beg to call your attention.

Should the money, or a part only,[66] have to be remitted to the
U. States, Mr Belmont might perhaps make some
arrangements[67] with the Government, authorizing the
authorities to take my drafts as a remittance and supposing
you only gained $1/2$ per cent clear, on a very large amount[68]
such as are constantly pouring into the coffers of the custom
house,[69] it would always be[70] a good business, and you would
incur no risk if you insured the full value of the gold in New
York or London.[71]

Concerning these same duties however it appears to be as
yet undecided whether the money which has been paid by

the merchants here under protest shall not be refunded
entirely[72] to them. I believe this question to be now under
the consideration of the Home Authorities, and if I could get
the information *before* any other party here of the decision of
the American Government I might probably be able to make
a very good operation by purchasing the interests of some of
the merchants at a discount of 25 or even 50 per cent. And
even should you not wish to speculate in a similar manner
upon any new information, it would be of great importance
to know beforehand what is likely to be done on this head,
for supposing the duties should be refunded, a large amount
of coin would be suddenly thrown into circulation and the
price of gold dust would immediately rise.[73]

I do not write to Mr Belmont on the subject[74] because, not
being personally acquainted with him, he might probably
pay no attention to any of my suggestions. Besides he was in
correspondence previous to my arrival in this country with
Mr Samuel Ward, formerly of the house[75] of Prime, Ward &
Co., and who is at present established here as a merchant
and banker, and anything I might suggest to him as likely to
suit you, he might order Mr Ward to do on his own
account.[76]

It is impossible for me to enumerate[77] all the different[78]
operations I might with great advantage undertake for your
account as[79] the face of affairs alters so rapidly here,[80] that
they would not allow of my awaiting your reply. If however
you feel disposed to invest in me the sufficient powers, I am
confident you will later have no cause to repent having done
so. At the same time, it is impossible for us[81] to foresee
whether by the time I receive your reply, the same facility
will exist of employing the[82] money to advantage.

Concerning the exchanges,[83] nothing is fixed until after the
arrival of the steamer, when I shall again have the honour to
address you.

I beg to remain, gentlemen,[84] your obedient servant


B. Davidson

[postscript] I beg most particularly to call your attention to the plan of issuing notes, if practicable it would entirely obviate the necessity of sending coin from Mexico and likewise be of most use to the country.

The system of working the gold by means of quicksilver is gradually giving ground and I think that in a years time very little gold will be extracted otherwise than by machinery.[85]


## VARIANTS

1. "mail" for "steamer"
2. "been" omitted
3. "portion" for "number"
4. "would" for "could"
5. "of the Country" added
6. "nor" for "or"
7. "majority" for "greater number"
8. "it will be very shortly"
9. "the best" for "a most profitable"
10. original illegible [Sablovifères?]
11. "already" omitted
12. "here established"
13. "however never will"
14. "alteration" for "change"
15. "can" added
16. "and" omitted
17. "actually" omitted
18. "for the disembarkation of" for "to disembark the"

19. "about" omitted
20. "in every direction" omitted
21. "speculating" for "speculators"
22. "it" for "this"
23. "hold" for "maintain"
24. "to" for "and"
25. 1000 feet
26. "so" added
27. "is" omitted
28. "take" for "receive"
29. "Le" omitted
30. "of" for "for"
31. "in general" omitted
32. "and storing" for "storing etc."
33. "upon" for "on"
34. "for" for "forward to"
35. "easy" for "an easy matter"
36. "here" added
37. "French" corrected to "English"
38. "so" omitted
39. "so" for "too"
40. "from France" omitted
41. "that are" omitted
42. "that" for "what can be"
43. "and sale" omitted
44. "about" omitted
45. "the" for "per"
46. "ounces paper current as a" for "ounce ... in"
47. "the oz." added
48. "relative value" for "price"
49. "there is" added
50. "is" omitted
51. "safe and advantageous"
52. "necessary" for "most essential"
53. "coinage" for "manufacture"
54. "then" omitted

55. "in" for "on"
56. "whether" added
57. "judge" for "decide"
58. "viz." for "namely"
59. "at a fixed exchange" omitted.
60. "place" for "instance"
61. "1" for "one"
62. "and" added
63. "the" added
64. "chiefly" added
65. "only accepting . . . custom house" omitted
66. "a part of this money" for "the money . . . only"
67. "arrangement" for "arrangements"
68. "amounts" for "amount"
69. "such as . . . house" omitted
70. "be always" for "always be"
71. "no risk would be incurred if the Gold were insured to its full value in New York" for "you would . . . London"
72. "entirely" omitted
73 "it would be very advantageous for several reasons" for "I might probably . . . rise"
74. "these subjects" for "the subject"
75. "firm" for "house"
76. "and banker . . . account" omitted
77. "to you" added
78. "various" for "different"
79. "as" omitted
80. "here" omitted
81. "me" for "us"
82. "the" omitted
83. "exchange" for "exchanges"
84. "I have the honor to remain" for "I beg . . . gentlemen"
85. postscript "(I beg . . . machinery)" omitted in original

From: John Luck, San Francisco

To: N M Rothschild & Sons, London

On: 29 April 1852

Gentlemen,

By the mail of the 17th inst., I had but the single pleasure of announcing my arrival, to which letter I beg reference, for an explanation of the combined causes which prevented my reaching here at an earlier period.

In conformity with your instructions I will now proceed to communicate such information as I have been enabled to gather relative to your business here, and which I trust may be useful to you; promising that Messrs Davidson and May received me with great frankness and kindness. There has not been the least hesitation in these gentlemen to answer any questions I have asked, nor the slightest withholding of any books, papers, etc., by or through which I could acquire a knowledge of their mode of conducting your affairs, differing as the system does from that which occupies the same class in London and other European cities. First with regard to the deposits, these are really of comparatively much less moment than anticipated, inasmuch as they are principally composed of small quantities of gold dust left by miners, who in most cases leave it but a short time, and either take bills for the amount on London, Paris, etc., whence the profit arises from the exchange, or they are charged the $\frac{1}{2}$% per month on withdrawal, for the safekeeping. This commission is stated to be generally paid in gold dust and amounts to but a few dollars at a time when it is added to the common stock, thus increasing the bulk of gold and decreasing in the same proportion the price given for the whole, a system adopted to render the accounts less complicated, though not quite so clear as they would be were the transactions shown. There are several such small deposits now in the vault wrapped up and sealed by the parties who left them, which when called for will be treated in the same way, viz. bought at the current rate, a bill of exchange given, or the 1/2% per month charged, the rate of exchange differing perhaps in as many instances as there are customers dealing. The great bulk of the gold remitted to you is purchased at Sacramento City by the agents of Messrs Davidson & May, and forwarded here or accumulated, that which is brought here being only from the miners who bring their small parcels with them, and who are on their route homewards. The prospects of supply from both the northern and southern districts are very good, and the sources almost unexhaustible, that is for many years, the northern producing at present almost six to two against the southern which is of the quartz formation. Business however is at San Francisco much divided in regard to gold dust purchases, there being several bankers who draw on London etc., and numbers of merchant jewellers etc., who all have notices of "Gold Dust & Coin Wanted" stuck on their premises. That which is bought by Messrs D&M is I think generally from European miners, the Americans preferring their own people, these latter are exceedingly jealous, as well as exceedingly knowing, and although it appears from the return of gold shipped from this place that some houses send the largest quantity, from what I can learn their business is done at a very cutting rate and for the purpose of swelling the figures; at the same time it is generally believed that Messrs D&M do the safest and most profitable.

Of the seven lots of property in Clay Street, two and a half have been sold and realized $6.862.50m the remaining portions are let and bring $1,100 per month and which if sold would I should think fetch $13 or $14,000. This sum looks rather small compared with the monthly letting, but the houses being of wood are only considered worth a few months purchase. This lot cost $19,000 altogether and up to January of this year there have been received for rent etc. $17,000, therefore were the remainder sold now, a very good profit would be shewn.

The house in California Street does not make so good a figure, as it cost originally with freight charges etc. about $14,000, and although $8,000 have been received for rent etc., it was sold for $6,000 and these to be paid at $200 a month, the last payment of which extending to January 1854.

The Montgomery Street house is really first rate property. It stands in one of the best situations in the city, and being at the corner of two streets possesses great advantages especially as it is perfectly fire proof. These premises have cost you say $50,000, but I should think they are worth that sum at the present time — premises such as these and in situations very inferior, is from $800 to $1,000 a month. I have heard that the house next door belonging to Messrs Argenti & Co. cost them $70,000 and the one adjoining them cost $80,000 but then they have stories above.

In Battery Street is erected one of the iron houses of Messrs Drusina & Co. At present this portion of the city is not very popular owing to the last fire having swept off so much of the property in the neighborhood. It is built on piles and is considered only equal to one made of wood and therefore unsafe; a portion of it however is just let temporarily at $200 per month. I believe when Battery Street is completed and more occupied with buildings this store will become valuable and I would suggest that as it is very large and capable of receiving a great amount of merchandizes, that the underneath part be filled up with sand, and the exterior built up with bricks. The filling in might be done by the contractors who are employed by the town council to fill in to complete the street, and the brickwork also might be contracted for, thus rendering the expense for the whole much less. It would then be a fire proof building and rentable at $800 or $1,000 a month, at all events it would make it a saleable property, and if the Custom House should be raised in the immediate vicinity, which I understand is likely of be the case, I believe it would then bring a sum equal to, and possibly more than that which it would cost you up to the very time of sale.

The three iron houses sold to Mr Griffing for $52,000 etc. for security of payment of principal and interest Mr Davidson holds a mortgage, I think very good sale, that is, a safe one. I have made several enquiries relative to Mr Griffing and have spoken to him casually three or four times, and from what I can learn and observe he is a perfectly honest man and will fulfill his engagements strictly, at least in this particular. Although Mr Griffing was fortunate enough to escape the fire, he was not so in regard to the rains, which loosened the rock that bounds his premises at the back. In December last a slip took place and destroyed the roofs and two of the bodies of the three stores, these he has been obliged to rebuild and refit at a great expense; they are now very substantial, he having put up the backs, front and sides of the stores. He has only paid at present $5,000 of the capital debt and a new arrangement has been entered into between Mr Davidson and himself as to the future payments and interest particulars of which are to be forwarded to you in course. The remainder of the iron houses or rather the iron which composed them is now only worth the price of old iron, and the demand for such material being very small it will realize but a meagre sum, a portion was disposed of a few days since at $115 a ton.

A person I have met here and whom I knew in England is acquainted with Mr Adams. Mr A never lived at San

Francisco but was a resident of Stockton. In the course of
conversation with my friend and in answer to a question he
said, "Well, Mr Adams is a very good sort of man, but his
success in trade has made him not only suspicious of every
body with whom he does business, but I think has a little
cracked him". Messrs Davidson & May admit having charged
Mr Adams the ½% commission, but the transaction was of a
very limited character, the great portion of the gold dust
being taken away by him as he said he could make more of
it at Philadelphia.

Of the bill for f25,000 on Messrs Hartog & Co. of Antwerp
it would seem that Mr Davidson was tempted to give it on
the strength of a letter of introduction by some friend, and
his belief that Mr Hartog was rich and would not repudiate
or dishonor his son's signature. The bill was for the purpose
of purchasing a vessel to go into the Chili trade which at the
time was apparently good, but the speculation turned out a
failure; but of $4,464 the amount of the bill, $500 only have
been received, and which I fear will be the total, for the
young man is about here, and without any visible means of
existence. Messrs D&M regret however that Mr Lambert did
not accept the terms offered by Mr Hartog senr., viz. to take
so much at certain periods, as they believe then the whole
amount would have been liquidated.

The £710.10 on Lindsay & Co. of London came before them
in the same questionable shape, and the [ ...] bond which
was taken as security was very insufficient to cover them.
The ship with her cargo of coals was seized on her return
and sold. At the time of sale, coals were at the lowest
possible price, and what with duty on the cargo, the
expenses of seizure but [for "by"?] the authorities, the wages
of the crew and all other charges incidental to such a case,
the balance received by Messrs D&M viz. $670 is perhaps all
that will ever be got out of this transaction, for Mr Blood the

delinquent is still in San Francisco, and not worth suing for a dollar.

A false statement seems to have been made by the person who applied to the Paris House about the purchase of French Rentes. This transaction on the books of Messrs D& M is perfectly clear; according to Mr May the person on his first appearance asked for some advice to purchase the Rentes he required, this he at length consented to do. The gold dust was brought [for "bought"?] at the current rate of the day, and a bill given for f21,538.85 the equivalent of $4,397.70 @ 4.90. There was no commission charged at all, and the 3¹/₂% spoken of must have been some calculation made by the man himself between the price of the dust and the exchange for the bill. Some annoyance may have been felt by this individual for it appears that about two days after he applied again, and asked if he could have his money returned when he would give up the bill, which was of course refused.

The expenses of the establishment are very modest indeed compared with many others and I think certainly not more than you would readily admit were necessary for carrying on the business; the personal expenses and mode of living, anything but extravagant, and the style of residence such as you would be rather surprised at, were you to see it. And now, gentlemen, whatever would have been the amount of business, and what the result, had more vigor and spirit been inspired, and a larger intercommunication with merchants taken place in the first place, I cannot say, but possibly at this time you might have had half the transactions of the city. This however required caution and the exercise of that caution was perhaps the course preferred, which if not the most advantageous, may have been the most prudent. Of course I cannot tell what private speculations may have been entered into either by Mr Davidson or Mr May, or whether

they are enriched by such means, but they both seem much
hurt and annoyed by the "suspicions" which they say are
contained in almost every letter they receive from London.
Ever since I have been in California, I have endeavored
studiously to become acquainted with such particulars as
bore most directly on your interest, and I have done so in
that way which would give the least offense to Messrs D&M,
or create the least suspicion as to the real object of my visit,
and I most candidly offer you this opinion—that these
gentlemen are conducting your affairs both well and
faithfully, for though some matters have come before you of
a suspicious character by report and otherwise, and there are
apparent discrepancies in the detail of the accounts rendered
to you, I sincerely believe the whole is carried on truly and
honestly.

The city of San Francisco is certainly a most extraordinary
place, and its commerce of a most extraordinary nature. Its
liability to destruction by fire excites a constant dread but
this will I think be allayed in a few years, as brick and
fireproof buildings are being raised in many parts of the
town. You would not be surprized at the consternation
produced by any alarm of fire, neither would you at the
destruction caused by it; for the majority of the houses being
of wood, and the streets and paths of the same material,
conflagration spreads like wildfire, and at once involves the
destruction of the whole place. In commercial business
fortunes are sometimes made in a twelvemonth, and lost
again in an hour by fire. Consignments of goods occasionally
realize enormous profits, for though they may arrive when
the markets are against them, a very short time will make
their value 30, 40, 50 and often 100 per cent beyond their
invoice prices. Whether such a state of things will continue
is perhaps questionable, for as the place gets more settled, a
more regular state of things is to be expected and prices
more generalized; but it will require time with a population

so rapidly increasing before any real steadiness in prices can exist.

The laws and regulations governing the State, County and City are at present very arbitrary and oppressive and all licences, duties and rates to which foreigners are especially subject, most vile and exorbitant, and liable to be made more so, whenever it suits the caprice or prejudice of the authorities. Every thing in the shape of living, such as food, clothing, furniture etc., excessively dear, and for hire and labor $5 a day is a modest sum; this latter possibly would soon be reduced by the immigration of thousands of Chinese, who arrive in large numbers weekly; but the Americans are getting very jealous of this fact, and are endeavouring to procure some legislative measure, by which the introduction of Asiatic foreigners shall be prohibited.

All those mining companies on [for "in"?] the Mariposa district which was said to belong to Colonel Freemont, must from all that we hear, turn out complete failures. There is no settled or established legal right in Colonel Freemont to grant a lease for any portion or sector of the land, and two or three agents of companies, who have been up the country to take possession of their claims, have been most unceremoniously turned away by the resident miners, who claim the right by custom and usage and consequently retain and hold possession.

The news we have received here lately from Australia relative to gold finding is extremely favorable, especially from the Port Philip district. In a letter shown me a few days since at the counting house by a gentleman who had received it from his father, were several instances of very large quantities having been collected and the parties named who found them. A captain of a vessel also who has just arrived, told me that it was being found very abundantly and was selling for from 63/- to 64/- an ounce, but that money was very scarce, and the seller could only obtain about 45/- per oz advance on it. The captain wanted $10 or $15,000 drawn direct on Australia and there is a constant enquiry for such bills here, and Messrs D&M think there might be a very good business between the two places, and I should think a moderately large shipment of sovereigns from England at an early period would realize a very excellent if not a very large profit. American coin will not [ . . . ] in Australia except to remit to San Francisco; and this might be worth while, if it were to be had at a discount of 3 or 4%.

In conclusion of this long letter I have only to say that I think my further stay here not requisite to you, inasmuch as I have given a full report of those matters which you required of me. This I trust will be as satisfactory as I believe it true and as there is [a] ship likely to sail for Australia in about a fortnight from this time, I shall avail myself of the opportunity to reach Port Philip where I hope to have the honor and pleasure of receiving your kind and continued commands.

I have the honor to remain, gentlemen, your most obedient servant

John Luck

[postscript] Since writing the above your favor of the 1st ult. has come to hand, the contents of which I beg leave to note.

# Index

Notes are included only when they contain information not in the text.

## — Index —

www.ingramcontent.com/pod-product-compliance
Lightning Source LLC
Chambersburg PA
CBHW080928100426
42812CB00007B/2406
*9781606180549*